T0400197

Mining

Dedicated to the memory of my father,
John Sydney Grainge Biggs

Mining
Why It's Essential for a Sustainable Future

Timothy Biggs

polity

First published in 2025 by Polity Press

Polity Press
65 Bridge Street
Cambridge CB2 1UR, UK

Polity Press
111 River Street
Hoboken, NJ 07030, USA

ISBN-13: 978-1-5095-6749-2
ISBN-13: 978-1-5095-6750-8 (pb)

A catalogue record for this book is available from the British Library.
Library of Congress Control Number: 2024950439

Typeset in 11 on 13 pt Sabon
by Cheshire Typesetting Ltd, Cuddington, Cheshire
Printed and bound in Great Britain by Ashford Colour Ltd

For further information on Polity, visit our website:
politybooks.com

Contents

About the author

Timothy Biggs is an adviser to the mining industry, based in London. Tim works directly with mining companies, as well as with mining-focused investment banks, risk consultants, investment funds and research and intelligence companies. He is Professor of Practice in Management and Mining at the Camborne School of Mines, part of the University of Exeter.

Tim was previously a long-serving partner of Deloitte, firstly in Australia and then in the UK. He worked with mining clients across the globe for many years, and for seven years was Deloitte's UK and EMEA (Europe, Middle East and Africa) Mining & Metals Leader. He was also the firm's global Lead Client Service Partner for a FTSE 100 major diversified mining company. Other clients included listed gold, copper and iron ore miners, as well as oil and gas, manufacturing and energy companies.

In these capacities Tim has spent over twenty-five years visiting mines, mine developments and mining company offices all over the world – more than thirty-five countries in all – significantly in Australia, Canada, South Africa, Chile, Brazil and Russia.

Tim has been a regular presenter at conferences, universities and industry groups, usually on topics related to the essential nature of mining, but also on getting to grips with its 'business side'.

Tim is married with four adult children. When not advising, talking or writing, he is involved with his church and enjoys reading, running, watching rugby and playing jazz and blues on the piano.

Foreword by Mark Cutifani CBE

While natural products and tools have been used to support life since animals with opposable thumbs first walked the earth, it is remarkable how little most people today understand mining, the use of its products and its importance to the future of people and planet.

The Ancient Greeks described the nature of the world in terms of earth, water, air and fire. In capturing the essence of our worldly existence, they understood that the products of the earth are fundamental to us. Yet in today's world the products of 'Mother Earth' are rarely seen or understood for what they are: the building blocks of sustainable civilization. While such a claim may seem wildly dramatic, when one considers the breadth of use of the products of modern mining, it is hard to understand why its provenance is not better understood, nor the lacklustre public perception of the industry. Or maybe it simply comes down to the fact that we assume too much knowledge in broader society for what is seen by those working in the industry as being self-evident.

In a world preoccupied with news and entertainment in all its forms, filling our moments with new information, we sometimes forget to step back and reflect on how the world works, what makes it work and how we could all help make it work better. In a conversation about the nature of things we might start with a simple observation, 'If it isn't grown, it is mined.' To unpack this simple statement, one has to explore the very nature of life, how we feed ourselves, how we provide shelter and warmth in our local communities and where the materials come from that support everything we do and everything we are.

I first met Tim Biggs when he was a partner at Deloitte and involved in our business at Anglo American. Tim would be part of our sustainability conversations and saw first-hand our community work, our engagement with a broader group of social stakeholders and the conversations around why mining is so important to society. We are kindred spirits in our view that the industry has not done itself justice in telling our story, both the good and the bad, and that we need to further improve, if only to promote a proper debate on how we can do better in that broader social context. And to his great credit, Tim has taken on the responsibility to help right that weakness, and his book provides a practical starting point for a much broader and more constructive conversation around the role of mining in society.

While Tim's journey of understanding in the world of mining does not pretend to be an examination of mining and its philosophical context, it is something more important and far more practical. Helping people understand mining's context in today's world is all about making sure we support those activities that are critical to ensure the sustainability of life and our planet as we know it. The products of mining ensure we have clean air to breathe, clean water to drink, food and shelter, energy . . . in fact everything we need to support us in our daily lives, and within its products we have the raw materials necessary to ensure the planet will remain both home and provider for future generations.

Tim tells the story of mining in an accessible way. He focuses on and explains how mining is important and impacts our daily lives. In explaining the issues that mining companies navigate, he helps people understand the nature of mining in society. He also doesn't let miners off the hook by making excuses for bad performance or bad behaviours. In the end, it is clear, we all have responsibilities to ensure our activities are sustainable and in everyone's best interests.

As we reflect on the 'worldly philosophers' and their inherent belief that the earth is part of a greater whole, maybe we are also helping people understand that the role of responsible mining is key to ensuring the sustainability of the world as we know it.

And as Tim so eloquently states it, 'Mining is essential for a sustainable future.'

Mark is the Chairman of Vale Base Metals and a director of TotalEnergies. He was formerly CEO of Anglo American plc and before that CEO of AngloGold Ashanti. Mark is a globally respected mining industry veteran with a career spanning almost fifty years. He is a tireless advocate for a more sustainable mining industry.

1

Mining is essential

The term 'essential' is very over-used. Everything is essential and thus nothing really is. The word has been robbed of its urgency and its ability to hasten action. So you might think it unwise for me to write a book with 'essential' in the title.

But I have done so, firstly, because it's a statement of fact; secondly, because not enough of us understand *why* mining is essential; and, thirdly, because mining is *so* essential that a sustainable future would be impossible without it.

This book seeks to explain why.

~~~~~~~~~~

A few years ago, I gave a presentation to a group of science and engineering students at a global 'top five' London university entitled 'Why Mining Is Essential'. Despite initial scepticism on the part of the audience, by the end I had won at least some of them over. Many students came to talk with me afterwards, and what became clear was a lack of understanding of, and perhaps a lack of appreciation for, the mining sector. At the time, this lack of understanding surprised me. Unfortunately, the
~~~~~~~~~~

growth of movements calling for outright bans on mineral extraction, and demands that mining and oil and gas companies should be banned from university campuses, makes me realize that perhaps the lack of understanding that leads to such calls has been present for some time. This concerned me: if there's so little knowledge about mining among a group made up of intelligent and enquiring minds like this, it's reasonable to assume the knowledge gap is at least as large in the broader community.

~~~~~~~~~~

Imagine for a moment that you are one of those young, intelligent, idealistic university students. You care about sustainability, you want to protect the environment, you have an interest in social justice and you are an avid consumer of all that social media provides.

Imagine further that one day you are alerted, via a message on your smartphone, to an environmental protest which is to take place in central London, outside the annual general meeting of a large mining company. As some of your friends tell you that mining companies are inherently bad, in fact almost as bad as those even-more-evil oil and gas companies, you decide to go along.

You fortify yourself with a warming cup of tea, put your phone on the charger while you take a quick shower and get ready to go. Then you get the lift down from the apartment where you live and head towards the Underground. You arrive at the station to discover the trains are delayed, so you quickly use your phone to book an Uber, or some other app-based car service, and it arrives a few minutes later. You hop in and journey towards the AGM to join your mates decrying those who extract stuff from under the ground.

Well, you should use the time stuck in London's traffic to make a quick reassessment of your plans for the day, because nothing you have done, nothing you are currently doing and nothing you will do would be possible without mining. Nothing.
~~~~~~~~~~

Let me give you a quick overview of why I say that, and why I say it so emphatically.

The modern apartment building you live in is held up by steel, manufactured from iron ore, metallurgical coal and various other metals and minerals. The windows of your apartment building are made from glass, which is basically sand, some limestone and soda ash. The electricity you take for granted to keep the lights on, power your appliances and charge up all your devices, including your smartphone, gets to and through your apartment on copper cables and wires, miles and miles of copper. The water you used for your cup of tea and to have a shower runs around your building in copper pipes. Your shower will be hot, almost certainly, because of your gas-fired boiler. Your boiler is made from steel, copper and various other metals. The gas gets to you along pipes which are variously made from steel, copper or high-strength plastics.

All of these things are the product of mining (other than the plastic pipes, which are a product of the oil and gas industry).

Your smartphone, the source of all your communication, all your information, indeed an indispensable part of your existence, is made entirely from minerals. The case is aluminium, which starts out as bauxite; the innards are all sorts of different metals but all of them from under the ground; the touchscreen is glass and a combination of rare earth elements. The battery contains lithium, nickel and probably cobalt. There will also be, relatively speaking, quite a bit of copper. The only parts that aren't metal are plastic, which, as we noted above, comes from oil. As oil is a mineral, the 'made entirely from minerals' statement stands.

But the reliance upon metals for your communication goes well beyond the device itself. The social media messages you send and receive on your smartphone, including the one you received about the protest against the mining company, are sent via enormous IT server farms, or 'data centres', operated by the major technology companies.

These server farms use astonishing amounts of electricity. They also require yet more copper for the thousands of miles of cables and wiring, steel for the buildings and all manner of other metals for the servers themselves, the controls, the power systems and the cooling systems. The huge, complex infrastructure required to make social media and 'cloud computing' possible depends upon an enormous quantity of metals and minerals.

And all of this is the product of mining.

The Underground, which you were intending to take, is built from vast quantities of steel and copper, with the infrastructure around it, and for that matter all the infrastructure of the city, made from even more massive quantities of concrete, brick, steel, copper, glass and all manner of other things, all of which came from either mining or quarrying. Quarrying, by the way, is just mining but less complicated. It's how we dig up things like aggregate (you may know it as gravel), sand, clay and rock.

The Uber car which you ended up taking is made from steel or aluminium, magnesium, nickel and copper. If it's an electric or a hybrid vehicle, then its battery is made from lithium, nickel, cobalt, manganese, graphite and copper. Depending upon what sort of car it is, it might have platinum and other platinum group metals in it as well. If the car is on-trend, it will probably have an unnecessarily enormous and distracting touchscreen which uses up even more hard-to-mine and even harder-to-process rare earth metals, just like your phone.

If it's an electric car, it will have got to you because it was charged up at a charging point, with electricity that travelled to the charging point and then into the car on even more miles of copper cables. That electricity will have been generated either with coal, gas, uranium or renewable energy such as solar or wind. Don't get too excited at the thought that the electricity came from renewable sources. Those wind turbines are made from massive quantities of steel, concrete, copper and a whole lot of obscure metals for their magnets and their control systems. Unless it was built here, the car itself will have got to the UK in the first place on a ship made entirely of steel.

All of these things are the product of mining.

Everything in the car that isn't metal, including the seat coverings, will be some sort of plastic, vinyl or artificial fibre, all of which start as oil. The only exception will be if the car has leather seats.

Is a pattern emerging? You will be able to join the demonstration only because of the products of mining, with some assistance from the products of oil and gas. The only thing you've done on this day that isn't directly connected to mining is use a teabag to make the cup of tea you had before leaving your apartment. And that teabag wouldn't have arrived at your house without planting, harvesting and manufacturing machinery, together with transport infrastructure, all made possible by the products of mining. So, actually, everything you've done this day is directly connected to mining.

You simply cannot get away from a dependence upon mining. Something to reflect on as you head to that AGM.

~~~~~~~~~~

My example above of our youthful, idealistic student isn't intended to be a cynical dig at students. As a part-time academic, I value their enthusiasm, their spirit of enquiry and their passion for causes they hold dear. They aren't alone in not knowing as much as they could about the source of just about everything except their food. To illustrate this point, let me share a brief anecdote from a real encounter with someone who was definitely not a young student.

My work travels once took me to Peru, a country that requires a visa. So I visited the Peruvian embassy's consular section to provide my fingerprints and collect my passport with the necessary visa inside. While waiting to be attended to, I was sitting next to an older man whose looks were, shall we say, a bit on the wild side. He soon struck up a
~~~~~~~~~~

conversation, and enquired why I was travelling to Peru. I told him I was going out there to visit a mining company. This news got him quite animated.

'Mines,' he exclaimed, 'they make a terrible mess, leave big scars on the landscape. Don't like them.' As I hope is evident by now, I was ready for this reaction, so I decided I would respond.

'Do you own a smartphone?' I asked him.

'Yes,' he replied in a neutral tone.

'Do you have electricity in your home?'

'Yes,' he responded again, not quite so neutral as he was trying to figure out where my line of questioning was heading.

'Do you own or ever travel in a car?' This time his response was a grunt; he had worked me out. 'Well,' I said, although I suspect I didn't need to, 'you couldn't do any of those things without copper, and a lot of it, and the only way to get it is to dig it up. So I'm sorry about the ugly holes in the ground, but if you want your phone, or your electricity, or your access to cars, or railways, or aeroplanes, or ships, or any electrical device or appliance ever invented, or computers or tablets or cameras or coffee machines, then you need copper. By the way, if you don't mind me asking, what do you do?'

'I'm a coffee buyer,' he told me, before turning the other way. As I love coffee, and purchase a lot of coffee beans, I couldn't be too grumpy with him. Hopefully, he will now have a little more sympathy for those who are busy producing vital and irreplaceable elements for the machines that enable baristas to make coffee from the beans he sources in South America.

The fact is, we all rely upon the products of mining for everything we do, every day, even if we don't realize it.

To that end, I have included, as an appendix to this book, a longer, but still by no means complete, list of examples of our reliance upon mining.

~~~~~~~~~~
~~~~~~~~~~

You may be asking, 'If mining is so essential, why do so few people know much about it?' Well, perhaps it boils down to its ubiquity. We all have a habit of taking things for granted. We particularly take for granted things that we see and use all the time, every day. So perhaps the mining industry's greatest problem in relation to our level of understanding of it is, perversely, the fact that its output is everywhere. The products of mining are so universal that it's very easy for people to go about their daily lives without stopping to wonder about the origins of everything they own and consume. And this isn't unreasonable. We shouldn't really expect people to ask 'I wonder where the copper in my television was dug up' every time they turn on the football.

The downside of this is a breezy ignorance when it comes to the importance of mining. Again, normal human behaviour means that there's little interest in stories about the origin of products we use every day, as long as they keep on coming without any drama. Conversely, sensational events, including tragic events involving the loss of life, are very newsworthy. This means there's almost no discussion or acknowledgement when mining goes right, but a lot of coverage, often graphic, on those rare occasions when mining goes spectacularly wrong. Think of the various mining-related disasters and environmental catastrophes that have made the headlines over the years. The full reporting of such incidents is both right and important. But the constant, boring, predictable production from mines never gets reported at all outside the financial or industry pages. The consequence is that, for most people, the only time the world of mining intrudes upon their consciences is when the story is overwhelmingly and tragically negative.

The 'only noticed when it's bad' factor is a major contributor to the poor image of mining. And it's difficult to change. The problem is exacerbated by the fact that the mining industry doesn't always help itself. For every example of activist overreach, there are unfortunately examples of demonstrably poor behaviour by mining companies.

Corruption; exploitation; environmental damage; waste; mismanagement: there have been many examples of all these things in the mining industry, and they need to be taken seriously and addressed too. I discuss this in more detail in chapter 7, 'Responsible and sustainable mining'.

There was a recent example in the clothing industry of alleged modern slavery in the supply chain of a 'fast-fashion' label. This generated significant negative press and much concern that our desire for very cheap clothing was blinding us to the exploitation of those involved in its manufacture. It was noted that it should be impossible, if they were paying people properly, for a company to sell shirts for three pounds and make a profit, and yet they did. Despite this, the company in question went on to increase both their revenue and their profitability. Similar supply-chain-related bad press over the years for the makers of certain well-known brands of sportswear, and particularly sports shoes, has done nothing to dent the popularity of these brands and the profits flowing to their owners. The manufacturing processes, employee welfare and factory locations of many of the world's most popular IT and communications devices, clothing brands, toys and even foodstuffs would probably not pass the ethics test if put under proper scrutiny, and yet they rarely are. And on those rare occasions when the issues are raised in the press, the response from consumers causes barely a ripple. Why is this? Why does corporate behaviour in general have such a minimal impact compared to the response to bad news on the mining front?

I think the answer lies in mining's remoteness from people's decision making compared to the retail industry, or indeed any industry which connects directly with consumers. When you buy a dress, or a pair of trainers, or a smartphone, you think about the look of the dress, the comfort of the trainers and the convenience, indeed necessity, of the smartphone. You don't think about the person behind the sewing machine or in the factory. Of course, everybody, when asked, will assert that they care very deeply about the plight of oppressed, effectively enslaved workers in the sweatshops and the factories. Unfortunately, the evidence from consumer behaviour shows that they don't think about it for long enough, or care about it deeply enough, as they keep buying the products anyway.

Mining, on the other hand, is a few steps removed from the consumer, so everyone who considers they have a social conscience can safely be affronted by the perceived bad behaviour of mining companies. They don't feel they have to worry that responding in a consistent way by, for example, boycotting their goods, would impact upon their lifestyles. I guess this is probably a good thing because boycotting all the goods that have input from the mining industry would be impossible unless you wish to live in a cave and eat by foraging for wild berries.

~~~~~~~~~~

Mining isn't just essential, however. It's essential for a sustainable future. It will seem counterintuitive for many people but the reality is that as we transition to a more sustainable world, with more renewable energy, we will have to do more mining. The need to dig things up will increase as we seek to reduce our footprint on this earth.

Electrification of vehicles, the substitution of electrically powered domestic heating for gas boilers and the electrification of previously gas-, diesel- or steam-powered processes in manufacturing will all require significant capacity augmentation and extension of electricity generation, transmission and distribution infrastructure. This will require a lot of copper, steel and other metals.

Renewable energy generation, by way of wind, solar or hydro, is without doubt much less polluting and more carbon-friendly at the point of generation. But it comes at a cost. One of those costs is the land that could otherwise be used for growing food, but that isn't a discussion for this book. The major cost we are interested in here is the metals processing and equipment manufacturing that is required to build the renewable energy installations themselves.
~~~~~~~~~~

These installations are entirely built from metals and minerals and the mining and quarrying required to obtain them is significant. There's no renewable energy without mining. Finally, batteries, both for vehicles and for large-scale storage of electricity generated by intermittent sources such as wind and solar, require large quantities of difficult-to-mine and difficult-to-process minerals. The energy transition isn't a cut-and-dried decision to stop using fossil fuels. It, like everything else, requires trade-offs and compromises.

There's no question that the energy transition will lead to more sustainable energy production, but it's very important to remember that this doesn't happen in a vacuum. Sustainable energy, and indeed sustainable transport, sustainable manufacturing, sustainable everything, will require a lot more metals and minerals, and that means a lot more mining.

2
Mining 101

But what is mining exactly? What does it involve? It's hard to appreciate something without understanding what it is and what it does, so in this chapter I explore the fundamentals of mining. We could call this 'Mining 101'.

In response to my request for a simple description of mining, a very senior leader at one of the world's largest mining companies once put it to me quite succinctly: 'We blow stuff up then pick it up and move it.' I don't wish this book to be overly technical, but I will get a little more technical than that. There are all sorts of ways to describe mining, and I'm going to do it very briefly, using the five stages of the mining life cycle. The first four are what happens on site: exploration, development, production and processing. I have then added shipping, given that there isn't much point mining anything if you don't send it somewhere.

Exploration

Before you can dig anything up, you need to find it. Exploration is the process of searching for and confirming the existence of mineral deposits which can then be economically extracted ('mined') and processed.

This is the work of mining geologists, also known as 'rock kickers'. They are great people, geologists. They are often a bit eccentric, perhaps a consequence of spending most of their working lives wandering through wild and sparsely inhabited landscapes, in all weathers, searching for unseen riches. Mining geologists tend to be happiest when clambering over remote outcrops or drilling exploration holes on faraway mountainsides, living their best lives with only the hot sun, drenching rain or freezing snow and a few like-minded team members for company.

For thousands of years, mineral deposits have been found by following outcrops and using visible markers such as these to indicate the presence (or optimistically assumed presence) of further resources under the earth. These days there is a lot of technology involved in mineral exploration, including techniques which help identify areas of interest without necessarily relying upon any surface indications at all. These include flying over the terrain with highly sensitive equipment that measures and records geological and magnetic 'anomalies' that can then be followed up with on-the-ground investigation.

To help in this process, the natural resources departments of many national and regional governments have, over the years, carried out quite comprehensive geological surveys using a variety of airborne and ground-level techniques. These have built up detailed maps of the geological formations of regions and indeed whole countries, which are often used as the starting point for the exploration of specific areas of interest.

The regulation of exploration varies from country to country, but in general terms, you can't just wander across the countryside looking for a deposit, and then, when you find it, start digging it up. You need permission from landowners to explore on their land. If it's public land, you may still need permission from government authorities or local indigenous groups.

Once you have the necessary approvals, you can start the often protracted work of focused exploration. Usually this will involve drilling. Lots of drilling.

Exploration drilling involves sending narrow pipes sometimes hundreds of metres into the ground and, in the process of doing so, capturing a 'core sample' of the ground you are drilling through, which is then brought back to the surface and analysed. This analysis will reveal exactly what you have drilled through, and whether there is any of the metal you are looking for. The limitation of exploration drilling is that the core sample shows you what is in the hole you have just drilled, but you have no way of knowing if the ground 1 metre to the side, or 10 metres to the side, will have the same characteristics, without drilling another hole.

So you do. If a hole comes up with encouraging results, the first thing you will probably do is drill another hole nearby, to see if the deposit, the 'orebody' that you have just drilled through, keeps going. And if it does, then, yes, you will probably drill yet another hole.

The limitation to exploration drilling is usually cost. Drilling is very expensive. A reasonable estimate is at least $200 per metre, so to drill a 500-metre exploration hole could cost you up to $100,000. And remember that at the exploration stage, there is no mine, so no production, so no revenue. You can only spend the money you have raised from investors, or borrowed from someone, to drill those holes.

The key point is: exploration is a very involved process, and invariably there will be a lot of dead ends and false dawns before a mining company get to the happy position of being satisfied that they have a deposit, or 'resource', worth developing.

Development

Once you have found your deposit, you need to confirm that it can be economically extracted. This is done through a feasibility study, sometimes multiple feasibility studies of increasing detail. If these lead to the conclusion that you should proceed towards development, you then need to get approvals, permits and licences from the relevant authorities and regulators. The most complex of these is usually the environmental approval. While you are waiting for all these approvals, you will probably be doing the detailed planning for your mine. Will your mine be open pit or underground? What infrastructure do you need? What processing plant should you build? What shipping is required? These are just a few of the many, many questions you will need to answer as you plan your mine. Once you have figured this all out and received the necessary approvals, you can start development.

Mine development will depend upon the type of mine you have decided you need. If it's an open-pit mine, you may need to prepare the ground above where the pit will be. In extreme situations, this can involve diverting rivers, removing the tops of mountains or stripping millions of tonnes of waste, all done, these days, in an environmentally sensitive way. If it's an underground mine, you will either need to excavate the tunnels, known as 'declines', to access the ore you want to mine, or, if that ore is sufficiently deep, you will need to sink a shaft, into which you will install what is effectively a giant lift ('elevator' for our American friends) to take workers and equipment down, and workers, equipment and ore up.

Then you need to build all the roads to access everything; pads (large, flat areas) to store stockpiles on; tanks for storing chemicals, fuel, acids and whatever other liquids you will need; and the processing plant to turn the ore you mine into the metals and minerals you want. The processing plant will need crushers, grinders, probably a water supply, and of course a power supply. You will also probably need to construct a tailings facility to take the waste from the processing plant. If you have to generate your own power, you will need to build a power plant too. You will also need offices, laboratories, potentially a water treatment plant, accommodation for workers if the mine is in a remote location, maybe a helipad or even an airstrip, maybe a railway line and associated infrastructure, no doubt even more roads, safety equipment, security arrangements, and I am sure there are things I have forgotten.

Once you have built everything, bought all the equipment you need and recruited a workforce, you are ready to commence production.

Production

It's the production part of the mining life cycle that most people think of when you say 'mining'. This is the bit where you actually 'dig stuff up'. There are all sorts of different mining methods, but you can divide them into two broad types of mine: open pit or underground.

Open-pit mining involves, as the name suggests, digging a hole (or 'pit') that progressively gets bigger. Having determined where the ore you want to process is, you first blast the rock to break it up sufficiently that you can move it. Then you dig up the rock with large mechanical shovels. You send the rock that doesn't contain ore to a waste dump, and you send the rock that contains ore that you want to process either straight to the processing plant, or to a stockpile to await processing. You do this with haul trucks, which for reasons of convention and safety are almost always yellow, and often the size of three- or four-storey houses. As this exercise goes on, the hole gets bigger and bigger and indeed can sometimes become enormous. The Super Pit, a gold mine at Kalgoorlie in Western Australia, is called 'super' not just because it's pretty amazing, but because of its gigantic size. And there are plenty of even bigger pits around the world of mining. Eventually, the pit may get to the point where you can't go any deeper. If the orebody continues beneath the bottom of the pit, and it's economic to extract it, you may decide to go underground and keep mining.

Underground mining has a similar basic premise. You blast the rock so you can extract it, and then you take the ore-bearing rock to the processing plant. To get to the deposit, you either drive down the declines I mentioned above, or if the ore is very deep underground, you go down the shaft, and then along declines, or 'drives', to access the rock face, or 'stope', you wish to mine. There are many permutations of the underground mining process but the idea is the same. You blast the ore-bearing rock and then take it to the surface for processing. Perhaps the straight-forward description of mining I quoted above is, after all, the best summary of the production process: 'We blow stuff up then pick it up and move it.'

Processing

The amount of processing involved will depend upon the particular metal or mineral you are mining. Coal, once mined, requires no further processing before being shipped out of the mine gate. You just blast it, dig it up, wash it, sort it and load it on a train.

Gold, at the other extreme, will usually come out of the ground as a tiny proportion of the gold-bearing ore that you dig up. Perhaps a couple of grams per tonne of ore, invisible to the naked eye. It's quite amazing to think that miners can economically extract something at 2 parts per million. To get the gold into a form you can sell you need to put the gold-bearing ore through a quite complex process involving a lot of crushing and grinding, a lot of water, a lot of chemicals and a lot of energy.

In between these two extremes are all the other minerals, including iron ore, which doesn't need much processing;

copper, which needs a bit more, including the use of a lot of acid; and rare earth elements, the processing of which involves a lot of water, ends up with some pretty toxic by-products and is, in general terms, pretty scary and quite polluting.

Processing is complex and is different for every metal. There are, however, a few things which most mines have in common. For a start, with hard-rock mining, the first stage of processing involves tipping the ore-bearing rock out of the haul trucks and into a crusher. This reduces the rock size sufficiently to then put it into grinders, which make the ore small enough that you can add water to form a slurry and then start the chemical part of the process.

The other thing most mines have in common is the tailings dam. These have, quite understandably, received bad press over the last few years, following some catastrophic dam failures. Unfortunately, they are also a necessary element of almost all mineral processing. A tailings dam is a facility into which you discharge the liquid waste from the processing circuit. Most tailings dams are designed so that over time the water will evaporate, leaving a dry waste that is much more stable and easy to manage. There are processing techniques being developed today that significantly reduce the amount of water required, which in turn reduces the volume of waste sent to tailings, but they are nonetheless an element of most mining operations.

Out of the end of your processing circuit will come metal in a form that you can ship to customers, or to another plant for further processing. Some common examples include: iron ore at a percentage concentration ready for shipping (for example '65 per cent iron ore'); copper concentrate or copper cathodes (which are thin sheets of pure copper); and gold in the form of doré bars, made up of about 80 per cent gold, ready for refining. Sometimes refining or further processing happens on site, but usually you need to ship your metal or mineral somewhere else. And, ultimately, you need to ship your product to market.

Shipping

The last part of the mining cycle is getting your product to market. This is a significant part of the process because mines are rarely close to their customers. Iron ore goes from places like Australia, Brazil and South Africa to Japan, China and South Korea; copper goes from Chile or Africa to Europe, China, the United States and in fact everywhere. Metallurgical coal goes from Australia or Indonesia to India and China.

So you need roads, railway lines, ports, ships and all the other infrastructure that goes along with these things. For precious metals, this is usually not such a big deal – even large gold mines will often take the smelted doré bars out by helicopter. But for bulk commodities like iron ore, it's a very significant part of the operational complexity and the cost of mining.

The mining cycle summary

Mining is, conceptually, a simple process. You find a deposit, develop the mine, mine the ore, process the metal and ship it for further processing or to the customer. In execution, however, it's very complex. At every stage of the mining life cycle there are decisions to be made, each of which can have multi-million- or indeed billion-dollar impacts on a mining company's financial outcomes. Given how long all this takes, there will no doubt be many surprises along the way. And, finally, there will be the impact of all manner of things outside your control, adding to complexity, cost and time.

This, then, is mining for metals and minerals: you find them, dig them up, process them and send them on their way.

~~~~~~~~~~
~~~~~~~~~~

Mining is, by its nature, global. You can't move a mine, so you must go to where the mines are. It's important to consider this as part of 'Mining 101', as understanding mining's global nature is important to appreciating what goes into running the world's major mining companies, which, after all, are the companies which produce the bulk of the metals we need for a sustainable future.

A mining company may be listed in London, or Toronto, or New York, or Johannesburg, or Sydney. They may be headquartered in London, or Santiago, or Johannesburg, or Vancouver, or Hong Kong, or Melbourne, or Moscow. Their mines may be in Chile, or Canada, or Kazakhstan, or West Africa, or Indonesia, or the United States, or Australia, or frankly anywhere. Then they may have processing plants in Finland, trading operations in Singapore and IT functions in Manila. Or they might have them all in completely different places again. The combinations are endless. Mining is a truly global industry.

The global nature of mining is down to a number of factors: the location of the founders and investors; the perceived best exchanges for fund raising for this class of asset; the most favourable places to attract the necessary head office talent; the best power, water, transport and infrastructure deals for operating processing plants; and, of course, the proximity to the 'immovable' location of the company's primary mining assets. As companies grow, they may be looking for particular types of asset, with the preferred geographical location being, to an extent, secondary. So, for example, they may end up with gold mines stretched from Ghana through Central Asia, across Canada and down to Australia, which all then need to be managed and run in a coherent and hopefully profitable manner.

There are a number of challenges to managing geographically dispersed assets and operations.

The obvious, and perhaps easiest to manage, challenges are the physical ones: distance; time zones; the cost, time and complexity of travel; the difficulties in meeting in person and visiting sites; and so on. Then there's the core business itself. Managing remote workforces is a skill. Then think of the complexities in managing supply chains with, potentially, equipment and consumables coming from one country to supply mines in another country (or countries), from which concentrate is shipped to processing plants in third countries, before metal is delivered to customers in yet further countries – all managed from a head office which is located somewhere else again.

But much more difficult challenges present themselves on the non-physical side of running a mining company. Global miners are at the mercy of multiple, often conflicting, requirements and obligations.

They have to deal with tax regimes which are inconsistent, horrendously complex and often impossible to interpret without recourse to tortuous judicial systems. They have to deal with multiple jurisdictions where government regulation and licensing regimes can also be inconsistent, stiflingly bureaucratic and subject to change without warning and often for domestic political reasons completely beyond the company's control. Different countries will have different rules on raising capital, dealing with retail investors and financial reporting. Global miners have to be alert to the impact of international treaties and trade agreements. They need to be careful not to fall foul of the impact of sanctions imposed by governments upon other governments, companies and individuals. They need to be alert to regulations around cross-border financing, anti-money laundering regulations and related reporting requirements. Inconsistent (that word again) and often contradictory approaches to the issue of climate change between different jurisdictions add to the regulatory burden. More general environmental regulations and accompanying sustainability reporting requirements constantly change and will be different in every country.

Global miners need to move people around, and that leads to difficulties with work permits, visas and employee taxes. If they want to minimize moving people around, they need to find ways to attract, train and retain talent locally. If they can't find it locally, they need to convince people to move to somewhere perhaps not overly attractive. And that gets you back to the global mobility issues of visas and work permits. Talking of doing things locally, every country has different obligations around dealing with and supporting the communities around mines. As financial and investor sentiment moves, so do exchange rates and thus the costs and rewards for doing international business. If these changes are significant enough, the rationale for companies' location decisions may disappear or need to be reconsidered. In some countries, competition between provincial or state governments leads to differing laws, regulations, taxes, licensing arrangements, environmental requirements, employment law and even non-standard transport infrastructure, all in the same nation (Australia being a prime example).

A final, but important, point, because it often gets forgotten, is the issue of culture. Dealing with different cultural norms, expectations and practices can be difficult. When mine managers are expatriates and the workforce is local, you do need to go to some effort to ensure the cultural divide stays as narrow as possible. Cross-cultural misunderstandings can contribute to a lot of the other issues discussed in this book, such as dealing with communities, dealing with governments and steering clear of bribery and corruption.

I always found it very helpful, on my various trips to operations around the world, to have with me someone who not only spoke the language but also was originally from the country we were visiting or had a strong affiliation with it. I also quickly learned that the notion that 'everyone speaks some English' is not necessarily true. In the Russian Far East, most of Francophone Africa and much of South America, once I was out of the main cities, it was very common to find that, no, they didn't understand a word I was saying. Then the colleague who spoke Russian, or French, or Spanish, or Portuguese wasn't just helpful but vital. Sometimes, I would find that a person I was dealing with did speak some English but was embarrassed because they didn't think it was good enough. I have a 100 per cent success rate at putting such people at ease and getting them to talk, simply by observing to them that their English was 1,000 per cent better than my Russian, French, Spanish or Portuguese.

I could share many anecdotes on my travels to visit mines, but in this instance, as it's in the customs and norms of the way we break bread that cultural differences really come to the fore, I will share a restaurant-related story from my time in the cold and weary town of Blagoveshchensk in the Russian Far East. I have visited a few restaurants there which have, on the final pages of their menus, a list of items and what you will be charged if you break them. Thus, you might say, a breakages menu.

It's slightly concerning that the restaurants feel it necessary to codify such things. It's even more alarming when you realize what's on the list. It starts fairly innocuously, with glasses, plates, that sort of thing. All the items which could be broken by accident or through carelessness but nothing more. But then it goes on to chairs and lamps. I am not sure what they are expecting to happen in these restaurants. And then, at the bottom of the list, is the table. They feel they must let you know what you will have to pay if you destroy their table. I remember on one occasion my colleagues and I were very tempted to hang around until closing time in one of these establishments, just to see what sort of vodka-fuelled damage was inflicted upon the furniture.

The point of all this is that one of the often neglected issues with the global nature of mining is the need to be respectful to host cultures, be able to communicate appropriately with the locals and understand when they do things differently because that is the way they do them in their culture.

Nothing will endear an incoming mining company more to the local community and their future employees, suppliers and neighbours than a genuine and heartfelt effort to embrace the local culture. The consequences of not doing so can be a mine's lifetime of distrust, antagonism and worse.

You may be wondering at this point why mining companies bother being global. This is a fair question, and sometimes companies, including the very largest ones, decide to get a bit less global in order to better manage what they regard as the core assets of their business. There are many examples of companies demerging non-core asset classes into separately listed companies, selling off non-core businesses to other companies or deciding to exit certain jurisdictions altogether. And, frankly, you can't blame them.

But mining *is* global, and it will remain so. It will remain global because the demand for metals and minerals is global, and in most instances that demand is in a different location to the mineral deposits, the ultimate immovable assets.

To think of just two of the best examples. Firstly, the Pilbara, in north-western Australia, contains perhaps the richest and biggest deposits of iron ore in the world. Not counting the mine workers, this is a very sparsely populated part of the world. There are, without doubt, many times more kangaroos than people. But as an iron ore region, it's absolutely world class. So all that ore needs to be sent a long way to be turned into steel, mostly to China, South Korea or Japan. Then it needs to be sent even farther to be used where it's needed in manufacturing, infrastructure or construction.

Secondly, consider the Central African copper belt, located largely in the Democratic Republic of Congo. The DRC is, sadly, a country with a long history of instability, outright civil war, violence, lawlessness and general mayhem. It's again a long way from most copper end-users. But as a copper region, it's world class. So miners factor in the risks of operating there, take the necessary precautions and get on with the business of mining the ore the world demands.

Because we can't live without the metal that comes from mining, demand is such that distance, logistical complexities, the threat of violence, unhelpful and often ineffective governments, suffocating regulation and all the rest is not enough, and will never be enough, to stop its ongoing extraction.

~~~~~~~~~~

Finally, in Mining 101, we need to talk about safety. Safety is the number one priority for all respectable mining companies. Read the annual reports of most major mining houses and safety will be the first thing they mention. Any meeting involving miners these days begins with a 'safety share' where someone in the group shares an important safety-related message for the benefit of the group. Having participated in a lot of safety shares over the years, it's remarkable how often you learn something new, useful and potentially life-saving.

A mining company's focus should always be on producing metals and minerals profitably, definitely; producing them sustainably, of course; but before all else, producing them safely. It has not always been universally the case, but it's certainly the case now, and has been, for most companies, for a long time. Companies and management are rightly judged on their safety record. Lost-time injuries are usually an important element of any performance assessment and bonus calculation. A death on a mine site can end the tenure of a CEO, acknowledging that ultimate responsibility for safety rests with them.

It's also important to acknowledge that safety failures can impact far beyond the mine gate: a gas explosion can send noxious fumes across vast swathes of the surrounding landscape; a tailings dam collapse can inundate communities below the mine in lethal rivers of toxic mud, and, sadly, this has happened in recent memory; the release or spillage of waste from mines or processing plants into watercourses can poison the water supply for local communities and indeed those far downstream. Safe mining
~~~~~~~~~~

and processing is a matter of life and death not just for miners, but for everyone.

Implementing and maintaining a culture and environment of workplace safety is hard work and expensive. Mines are a safety nightmare: they often cover physically large areas; there's the potential for dangerously unstable ground; there are often lots of vertical or near-vertical drops; there's a lot of heavy moving machinery; they will use a cocktail of chemicals and acids; rock gets blasted with significant amounts of high explosives; there are processing plants with even more chemicals, moving parts and confined spaces; if the mine is underground, there are even more risks and dangers I haven't mentioned; and, of course, there are the many risks involving tailings dams, large bodies of often toxic liquid. And this list is just the start.

So when developing a mine, an understanding of physical safety risks and a plan to mitigate and manage them is essential. And again, like everything else with mining, this will take time, will need to be exhaustive, will cost a lot of money and will need to be developed in parallel with everything else. It therefore needs to be planned from the start. The cost, time and effort that are necessary to properly address safety, both in planning and ongoing monitoring once the mine is in production, aren't an option, and nor should we begrudge them. Safety must genuinely be of primary importance. I flag it up for two reasons. Firstly, because it must always be top of mind, and, secondly, because being so essential does not alter the fact that it's one more thing that adds to the complexity and difficulty of mining.

A truly safe mine also requires a 'top-down' safety culture. It's difficult for teams to take safety seriously if they perceive that management does not. And it needs to be ingrained. I once visited an underground gold mine in South Africa where the mine manager had just returned from a six-month exchange with an Australian gold mine. He observed how hard it was to build a safety culture from scratch. In the Australian mine, he had seen that safe work practices were so 'built in' that all you had to do to ensure the mine workers didn't go into unsafe parts of the mine was stretch some red-and-white safety tape across the tunnel. So if, for example, you had just blasted some rock but hadn't yet put in the roof bolts to make the area safe, some basic signage and the aforementioned stripy tape was all that was necessary. Everyone working on the mine knew that the safety tape meant there was a risk of rock falls and you mustn't proceed past it. So they just didn't. By contrast, in the South African mine, in the same circumstance, they had to put up strong wire grating to stop workers going into newly blasted areas. All the warnings in the world didn't stop them trying, because there was no ingrained culture of safety and the lure of maybe finding visible gold to be pocketed was stronger than the fear of being flattened by part of the roof detaching at an inopportune moment. To be fair, the company were working very hard at improving their safety culture but it was a long process because it involved changing attitudes. Such things take time, persistence and constant reinforcement. But the industry as a whole demonstrates that it can be done, and done successfully.

The primary reason, quite correctly, for the almost obsessive focus on safety in mining is because miners want to ensure no-one is hurt and no-one is killed on a mine or in a processing plant. But there are many other benefits to a strong safety culture and it's worth noting them, although I stress again that these are secondary to the goal of ensuring everyone goes home alive and uninjured at the end of their shift.

Firstly, any kind of serious accident or incident leading to lost-time injuries or deaths will result in the mine, or the relevant part of the mine, being shut down, probably for a long period. Investigations by both the company and regulators will take time. Where structural or operational changes are then required, the shutdown time can be very long. And we are talking here about months or even years, not weeks. The cost of rectification isn't the only problem. The big cost is lost revenue from lost production. Additionally, if the incident has led to the destruction of property and loss of life beyond the mine gate, the costs can become astronomical.

A shutdown, the loss of revenue and the associated costs may call into question the viability of the mine. If the mine is out of action for a long time, or even permanently, this will have a potentially life-changing impact on the workers who depend upon it for their wages. It will similarly impact upon the suppliers, contractors and others without tenure who may also rely upon it for their living. The impact will be exacerbated if the mine is (or was) the only significant source of economic activity in the area.

The reputational consequences of a poor safety record can also have an impact. It may result in suppliers not wanting to deal with the company, or mean that the recruitment or retention of staff becomes a problem. It may result in a more intrusive approach from host governments, in the form of regulatory oversight, the time taken for licence and permit renewals or just the general level of cooperation and support. A poor safety record may also impact upon a listed company's share price and lead to investor pressure for change, including changes in senior management, particularly if there's a view that the 'tone from the top' isn't adequate.

One of the complicating factors with safety is that dealing with it can be counterintuitive. The Queensland coalfields in north-eastern Australia are, in part, close enough to the coast that a lot of mine workers will drive a few hours each way at the beginning and end of their time at the mine. One of the mining companies in the region was concerned about the risk of accidents as this practice became more common, so they undertook a survey. The 'received wisdom' was that miners travelling home after ten days of twelve-hour shifts would be at significant risk of having driving accidents as they would be tired from all that hard work.

What they found was the opposite. Out on the mine site, they basically worked, ate and slept for ten days. Yes they worked hard, but they didn't 'play hard' – they just ate a lot of excellent food and slept uninterrupted sleep. They drove home to the Queensland coast frankly quite refreshed. As many of them had young families at home, they then arrived, after ten days away, to four days of 'not working', which was more exhausting than anything they did at the mine. They had ten days of absent parenting to make up for in four days, and their wives or husbands made sure they did their share of parenting in that four days. By the time they jumped in their cars and drove back to the mine, they were sleep-deprived and worn out. The accident risk was on the way to work, not going home. You can never be complacent when considering mine safety.

There can also, though, be quietly amusing elements, even in the very serious business of mine safety. One of the first mines I visited in the early days of my focus on the industry was the now closed Mount Leyshon gold mine in North Queensland. This mine was remarkable. It started out as a mountain and ended up as an open pit. Over the years of its life, a deposit the shape of an enormous egg was mined, producing over two million ounces of gold. But weirdly the thing I remember the most about visiting this mine was my introduction to the importance, in relation to mine safety, of adequate hydration. Apologies in advance for getting a bit biological.

In the men's bathroom, next to the trough was a laminated poster with a colour chart graduated from clear through to dark yellow. Above the chart was an instruction to check the colour of your pee. If it was below a certain point on the colour chart, there was a likelihood that you were dehydrated and you were instructed not to proceed into the mine itself until you had topped up on fluids. I found the simplicity and clarity of this safety procedure both impressive and effective.

Of course, there could be unintended consequences if, after consulting the chart, you decided you needed to quickly drink a litre of water. You would then find, during your trip through the mine, probably at the farthest point from being able to do anything about it, that you were stuck in the back of a small truck with unforgiving suspension, rattling along the bottom of the open pit, absolutely desperate to visit the loo!

But don't let my anecdote distract you from this key point: mining is risky, it can be dangerous and the primary focus must always be on mining safely.

~~~~~~~~~~

The purpose of this chapter was to introduce the fundamentals of mining, and then broaden things out with an understanding of a couple of the major issues for mining: the global nature of many mining ventures, and the overarching importance of safety. In the next couple of chapters, we will consider some of the key challenges facing mining companies ('miners') and, following that, having given you an understanding of what mining is and the difficulties miners face, we will turn to the essential nature of mining in a sustainable world.
~~~~~~~~~~

3
The operational challenges of mining

Miners face many operational challenges, and in this chapter I want to talk about a few of the key things that make the life of a mining company more difficult as they seek to produce the metals and minerals we all need. Let's start with 'the major capital projects budget curse'.

Whatever you think it will cost to develop a mine, double it, then perhaps add another 50 per cent. I appreciate that this is a flagrant generalization, but as many investors have found to their discomfort, probably not an unreasonable one. Why is this? Why has the tenure of many a mining CEO come to a premature end because of a massive capital project-related cost blowout? Why are miners so often at the mercy of the major capital projects budget curse?

As we have already noted, mining projects are almost without exception very long term. Once a miner has discovered a deposit, done a feasibility study and decided it's worth developing, there will be years, potentially a decade or two, before it comes into production. If it's a big deposit, this whole process may cost billions of dollars. The company won't get a cent of those billions back until the mine is in production and they have product to sell. This, by way of a short segue, is the reason even the largest mining companies will often sell down a stake in their mine developments to a third-party investor in order to spread both the cost and the risk.

So getting the budget right is of paramount importance, obviously. Why, then, do budgets so often turn out to be wrong?

There are many reasons, and I flag these up to help us understand more of the factors which contribute to the high-risk nature of the mining industry. Next time you are sitting comfortably in a city coffee shop, reading about the sorry tale of a far-away mining project costing billions more than originally budgeted, think about these issues before you murmur, 'Incompetent! They deserve to be sacked.' An alternative expression comes to mind: 'There but for the grace of God go I.'

The long-term nature of a mining project is the first difficulty working against accuracy in forecasting. In order to get investors or other funders to commit to a project, it will need a full-blown feasibility study, or 'bankable' feasibility study. You need this before you do just about anything, which means that you are talking very big numbers. And these very big numbers have to be plugged into your feasibility study years in advance.

Between the feasibility study and producing ore, the work is significant. On the external front, you need to obtain approvals, arrange infrastructure (roads, railways, power, water, etc.), engage with the local community and agree on how you will support them, and call for and award tenders for construction. On the internal front, you need to be developing detailed mine plans and plans for the processing plant and supporting infrastructure, while almost certainly doing more drilling, called 'infill drilling', to firm up your reserve and resource numbers, consider your workforce requirements and, of course, get a management team in place. Then you need to actually build everything. Finally, after a lot of testing, and no doubt a lot of rectification, you will be ready to operate your mine.

What all the elements in the preceding paragraph mean is that developing a mining project takes a very long time and involves many complex and interrelated elements. And in the time between starting the feasibility study, when you have to come up with a forecast cost, and finishing construction, when you know exactly what you've spent, the cost of everything will probably change, no doubt in an unhelpful direction. So this is the first difficulty. Construction contracts, service contracts, equipment purchase prices, wage rates, these will all no doubt be different by the time you get to the point of starting to build your mine and order your haul trucks. And when the prices change, just to make absolutely clear, they invariably go up, not down. Your capital project usually costs more than you planned. It absolutely never costs less.

During the mining boom years in Australia in the early 2000s, the number of haul trucks increased exponentially. As a consequence, not only did the cost of these trucks go through the roof, so did the cost of consumables and, most notably, tyres. Even back then, a haul truck tyre could cost you more than $50,000. But more of a problem was actually finding tyres to purchase. So acute did the shortage become that companies were digging up old, discarded tyres from where they had been placed, often as parts of retaining walls, or to shore up temporary power poles, or as safety barriers on haul roads, and pressing them back into service on the trucks themselves. It's reasonable to assume that this scenario wasn't planned for when the feasibility studies were being undertaken.

The second difficulty is that in addition to the costs increasing, the plans themselves will change. And invariably when plans change, it's because you need to add things, not take them away, so again the costs go up. For example, the geology of your orebody will without doubt throw up some surprises. Indeed the geology of your deposit will almost certainly not turn out to be quite what you originally planned for, regardless of how many holes you drill. Changes in ground conditions, geology and the metallurgy of the orebody will change the way you mine, precisely where you mine and, perhaps most costly of all, how you process the ore. The ore may be harder, or softer, or in some other metallurgical way different to what you anticipated, resulting in changes to the design of your processing plant. And I have not encountered many mines where any kind of design change led to a cost saving.

The third difficulty is regulatory change. There's a vast body of regulations that any miner needs to comply with in the development of a new mining project. These will cover environmental matters, such as emissions, water usage, noise, smell (yes, really), power usage, personnel matters, community engagement and more. When any of these change, particularly those with a direct mining impact such as emissions, water and power, it will probably require changes to the design of the project and thus to the cost. For example, the water you were planning to draw from a nearby river may be restricted or you may be prohibited from using it after an unanticipated change in environmental regulations. So you then need to build a 300-kilometre pipeline from the nearest ocean together with a massive desalination plant. Or the regional government may decide to increase your 'voluntary contribution' to local infrastructure, so you need to factor in the cost of another hospital. Who knows what might change, but it will certainly be more expensive than you budgeted for.

Making all this even more complicated is that a lot of mining companies contract out the entire development project to an external contractor. There are various ways this is done but most common are EPC (Engineer, Procure and Construct) contracts, also known as 'design and construct' contracts, or the even more outsourced EPCM (Engineering, Procurement and Construction Management) contracts. Although in theory this should result in greater efficiency and thus less cost because the contractor is skilled at building mines, it doesn't always work out that way. Usually this isn't because of rapacious greed or incompetence on the part of the contractor, but because the project isn't defined with sufficient precision in the contract, or there's an inadequate mechanism for dealing with design changes, and there are consequently disputes and overruns caused by unforeseen technical issues, planning or regulatory changes, procurement or construction delays, subcontractor disputes and other problems. At such times, mining companies often realize to their cost that they are contractually bound to either continue with the project or pay significant termination costs, even as the economics have fundamentally, negatively altered.

There are all sorts of things that can change. Ultimately, it comes down to the potentially toxic mix of long time periods, unforeseen changes in mine, plant and infrastructure plans, escalations in procurement costs, changes in economic or regulatory circumstances and inadequately drafted contracts.

This all costs money. Vast pots of money. And thus the major capital projects budget curse.

~~~~~~~~~~

A second vital factor in the success of a mining company is the quality of their management.

Mines are, as we have noted (repeatedly), very complex operations. They thus require a lot of management. Indeed the whole premise of this book, which is that mining isn't just essential but also complex, difficult and beset by all manner of factors outside the miners' direct control, should make the need for excellent management self-evident. I don't think anyone intentionally sets out to have bad management, but having good management isn't nearly as easy as you might assume.

Putting in place a management team for a mine development is hard work. You need to find the right mix of experience, wisdom, leadership capabilities, people skills and technical skills. You then need to find people with the attributes I list above who are willing to live in, or at least spend a significant amount of time in, the area where you are developing your mine. If your mine is in
~~~~~~~~~~

a remote location, and we have already observed that most of them are, then attracting good management will probably both be expensive and require you to promise high-quality facilities and amenities. Security may also be an issue. Attracting quality management to a gold mine in Mali, for example, may be problematic.

Many years ago, I attended a dinner where the guest of honour was Li Ka-shing, at the time the Chairman of Cheung Kong, the enormous and very successful Hong Kong-based conglomerate. After his speech, he took questions, and he was asked what his key success factors were for the operation of his many investments around the world. His answer was both simple and instructive. The number one thing, once he had bought an asset, was to have the right person running it. So he only bought assets once he was satisfied he would be able to put the right management on the ground. Having the right people to run his investments was as crucial to success as the quality of the underlying assets themselves. Or put another way, there was no point buying excellent assets if you didn't then have excellent people to run them.

Amusingly, he went on to note two things that were often forgotten about when assessing how easy it would be to convince people to relocate to manage assets. Remuneration, security, quality of housing, ability to take the family, availability of good schools and all the rest were always talked about. But two he thought were more important than they were given credit for were good air links and a convenient time zone. The dinner in question was in Adelaide, South Australia, where Cheung Kong had recently purchased the state's electricity distribution assets. Adelaide is a delightful city, but, back then, it had pretty hopeless international air links and an illogical and irritating half-hour time zone difference compared to the east coast of Australia. Perhaps he was having a gentle dig at his hosts?

So what else do mining companies need to consider in putting together an effective management team? It's probably worth considering in the context of the major elements of management effort. Of, course those major elements will differ depending upon the stage of the project. The skills and attributes of management will be different depending upon whether you are developing a project or running an operational mine.

In a development project, you will need a lot of technical management: geologists and mining engineers to work on your mine plan; construction engineers and construction managers to build everything; managers who are good at dealing with governments and regulators to smooth the way to permitting and licensing; and managers who understand people and communities to ensure you gain that elusive 'social licence to operate' with the communities around your project, and then you are able to recruit a suitable workforce to build everything and ultimately operate your mine.

Once you start operations, you then need people who can manage the operation; manage the workforce, including managing a workforce that may be culturally different to the managers' own experience; manage ongoing relations with external stakeholders, including communities, governments, suppliers and shareholders; and have the nous, the experience and perhaps the 'rat cunning' to keep the show on the road through all the ups and downs that come with running a mine.

I noted earlier, 'I don't think anyone intentionally sets out to have bad management', but, strangely enough, sometimes without realizing it, that happens. It could be described as the founder's curse, and applies as much to tech companies as to mining companies. Mining companies often start out as exploration companies, run by geologists. If they are successful, they will over time transition from being a small company with a tight-knit team doing what they love best, kicking rocks, to a development company dealing with all the complexities of building a mine, and ultimately being a producing mine beset with all the management issues of a fully fledged commercial operation. Being a successful geologist doesn't necessarily translate into being a successful CEO of an operating mining company. Indeed, history tells us that when highly skilled technical people turn their hand to management, the failure rate is pretty high. And thus the founder's curse. It must be a horrible job to be the chairman of such an enterprise when you have to tell the guy who started it all that he needs to step aside for the good of the company.

Of course, it's not just a problem when someone allows their technical prowess and perhaps technical pride to override sensible economic decision making. It can, and does, happen just as easily in reverse. Finance managers can stop listening to the techies, and make decisions using assumptions not based in the physical realities of the mine, or they have an obsession with costs without regard for value creation, or they chase market share without considering the risks, or aggressively growing sales means that words of caution from those who understand the technical side of the business can be ignored until it's too late. The embarrassing profit warning, the cash crunch or the quality problems often follow soon afterwards.

So, for a mining venture to be successful, it's important that there's the right balance and a good relationship between the business and technical sides of mine management.

Finding and retaining good management requires sustained effort, and one of the most important roles of the board of any mining company is ensuring their management team, starting with the CEO, is up to the task. We talk a lot these days about how important people are to the success of organizations. Well, managers are people too, and a management team will make or break your mining venture.

~~~~~~~~~~

A third challenge I want to flag up in this chapter is around selling a mine's production.
~~~~~~~~~~

There has been a lot of press in recent years about the manufacturers of electric vehicles, or, more specifically, the manufacturers of the batteries for electric vehicles, seeking to secure the supply of the necessary battery metals by signing long-term contracts with mining companies or commodity traders.

This makes sense. There's a lot of uncertainty around the supply of some of these metals, and taking one of the variables out of the supply chain equation seems like a prudent thing to do. Of course, a bit like hedging, the manufacturers may find themselves locked into agreements which are 'out of the money', as they say, if supply surges and prices drop. They may also find themselves locked into purchase agreements for materials that become surplus to requirements as a consequence of innovation, always a risk when operating at the cutting edge of technology. But such risks are normal. Any risk-mitigation strategy comes with an opportunity cost.

The risk of supply disruption, or of simply not being able to obtain the supply you need at all, is becoming very real in relation to many battery metals, rare earth elements and even some relatively widely used metals like cobalt, magnesium and tungsten. Where the vast majority of production and processing is in China, and where demand, including Chinese demand, has on occasion exceeded supply, it's not surprising that overseas processors and manufacturers have at times struggled to secure a supply of metal. This will become a material impediment to progress on many renewable energy projects in much the same way as the global shortage of microchips at the beginning of the 2020s had a drastic impact on motor vehicle production.

The issue noted above on the 'buy side' is, of course, mirrored on the 'sell side' and nicely highlights one of the other factors to be considered when thinking about or planning a mining project. To what extent should you agree to sell your production in advance? Does it make sense to enter into a supply agreement, known as an 'offtake agreement', with a third party? And what is the potential opportunity cost I mention above for the producer? The answer, of course, will largely be dependent upon the individual circumstances of the miner; their view of future demand, supply and thus prices; and their risk appetite.

There are a few other things for producers to consider, including concentration risk. A good example of this is concentration risk added to the risk of unexpected interruptions arising from geopolitical factors, best exemplified by the Australian iron ore trade. About 70 per cent of Australia's iron ore production goes to Chinese steel mills. Profits from iron ore account for about 90 per cent of the total profits of the major mining companies involved. So this means that roughly 63 per cent of the profits of these diversified global mining companies actually come from one overarching transaction: the sale of Australian iron ore to Chinese steel mills. That is a big 'concentration risk'. Now I appreciate that this risk is mitigated to an extent by the contention that China couldn't readily replace this volume quickly from somewhere else, but it's still a significant risk. And when you're dealing with an authoritarian state where the interests of the ruling party trump all else, you can never be sure you won't get caught up in unilateral actions which disrupt your sales and over which you have no control.

Ideally, from a certainty point of view, you could enter into something like a 'take or pay' contract, where a customer agrees to buy your output for a certain period of time, and has to pay you even if they don't take it. The downside of this is once again the opportunity cost. If the spot price goes up beyond the price in your take or pay contract, you are missing out.

Why is all this significant? It's significant because when planning a mining project, determining the right balance between having the security and certainty of a big customer (or customers), on the one hand, and opportunity cost and concentration risk, on the other hand, will be vital when assessing the economics of the project and may also have a major impact on the ability to obtain funding.

Yet another element in the complex matrix of risks, opportunities and uncertain outcomes that is inherent in planning a mining venture.

~~~~~~~~~~

The final challenge for miners in building and operating their mines is the management of risk.

Managing risk, for a mining company, is central to what they do. There are certain types of safety-related incidents that can be genuinely existential threats for a miner, and therefore serious work needs to go into mitigating and managing the risk of them happening in the first place. But there are many other risks, and indeed I use the word 'risk' so often throughout this book that it should already be clear how much risk there is in the mining business.

The first thing to note is that even with improvements in safety, advances in systems and processes and a degree of prudence in dealings with finances, supply chains, regulatory matters and much else, mining is still an inherently risky business. We aren't talking about risk elimination, we are talking about risk mitigation and risk management.

It's valid to note that a lot of mining companies make significant profits, and that mining is thus a 'high-risk, high-return' industry. The thing to remember, however, is that for all the established mining companies that are financially very successful, there are even more that fail, often because they don't manage their risks.

The risks that mining companies face are too numerous to fully outline, but what follows is a representative sample.

Commodity price movements. The main risk obviously is that the price you can obtain for what you are producing goes down. If, for example, a mining company
~~~~~~~~~~

find themselves over-hedged and under-producing, they may also find themselves in a position where there's risk attached to the commodity price going up.

Reduction in demand from a major customer or market. For example, if you sell all or most of your product to one particularly large Asian country, and then get caught up in the middle of a geopolitical squabble, that could cause you a problem.

Political instability or unhelpful changes in government. If you are operating in a historically stable country which undergoes a coup, an attempted coup, a period of prolonged civil and political unrest or, for whatever reason, elects a new government which has a dislike for the mining industry, or perhaps just a dislike for foreign-controlled mining companies, you could also have a problem, potentially a big problem.

Community conflict or unrest. Mines, as we have previously noted, often co-exist with local communities. The aim is always to have a productive and mutually beneficial relationship with local communities, but for a variety of reasons things don't always turn out that way, and you end up with conflict, disagreements and disputes. Unfortunately, this can extend to disruption, such as blockading the mine, violence and even, on rare occasions, armed conflict.

Uncertain and changing government regulation. We have already noted that obtaining all the necessary approvals, permits, licences and agreements from governments is time-consuming and expensive. There's also the risk that regulations and requirements change, sometimes in the middle of the process. Perhaps most often in relation to environmental approvals, it can be unclear what the requirements actually are, and what a potential mining project needs to demonstrate to get over the line.

Reserves and resources estimation. Ultimately, a mine is built to extract the reserves you have identified. Your feasibility studies, mine planning and financing are all built on the assumption that there's at least a certain amount of ore containing metal of a certain grade that you can economically dig up. If that turns out to be wrong, or materially overestimated, or perhaps there's a geological problem that wasn't identified, then your project may not be viable after all. And you may not find this out until you have spent a lot of money.

Mining methods. From very early in the process, and usually as part of a feasibility study, a project will determine the method of mining. Will the mine be open pit or underground? If underground, what particular mining method will be used? What form of access will be developed: declines or shafts? What form of haulage: conveyors, trucks, tramways or something else? There are many more questions that need to be answered, and, once answered, a vast amount of money, in many cases billions of dollars, is spent on developing the solutions. There's a risk that the chosen solution turns out to be suboptimal, or, worst of all, simply doesn't work in practice. Rectifying the situation may vastly reduce the profitability of the mine, or could make the whole enterprise uneconomic.

Inputs and input costs outside your control. This is actually a number of risks in one. Any mine relies upon things like a reliable power supply, a reliable source of water, fuel and a lot of materials, consumables and equipment to operate the mine. If availability goes down, or the price goes up, or both, your economics may no longer work the way you thought they would. If availability of vital inputs, such as energy, water, fuel or essential equipment, declines or stops, then mining may be disrupted or even halted. We are seeing the impact of an input-related risk coming to fruition at the moment with the increasingly precarious state of South Africa's energy network and its impact on the mining industry (and everyone else).

Uninsurable hazards and events. One of the consequences of the tailings dam disasters of the last few years and the response to them has been that it's now almost impossible to obtain insurance for this aspect of a mine's operations unless you have a verifiable and thorough risk mitigation and management strategy. This will probably include active management and constant physical monitoring. Mining companies need to ensure they have identified potentially uninsurable risks such that actions are taken to make them insurable, or the risk itself is managed such that companies can effectively self-insure.

Capital expenditure overruns. Mining projects cost a lot of money, and as we noted in relation to the major capital projects budget curse, they have a bad habit of running significantly over budget, both in cost and in time. This is a risk that takes a lot of managing, particularly if the economics of your project were tight in the first place.

Operational safety, incidents and accidents. This risk can relate to small but avoidable accidents which may result in damaged machinery, some lost production and nothing else all the way up to catastrophic accidents such as an underground gas explosion or a pit wall collapse. Whatever they are, miners need to have processes and procedures in place to manage the risks down to an acceptable level. Particularly when it comes to safety, every decent mining company is aiming for zero harm, and that requires constant management of the risk of serious incidents.

Labour disputes. Mining, in most cases, isn't the labour-intensive industry of many years ago. Mechanization, technological advances and modern mining methods have resulted in vastly more ore being mined with far fewer people. Unfortunately, however, there are still many instances of difficult labour relations between mining companies and their workforces, particularly where unions are involved. In fact, the reduction in workforce size has on occasion exacerbated this, with unions actively seeking to preserve their bargaining power in the face of a reduced, and at the same time more highly skilled, workforce. Whatever the cause, disruption to production and the ongoing difficulties caused by fractious labour relations are risks miners must deal with.

Corruption. We have talked about this risk previously. The risk will vary depending upon the jurisdiction and market that companies are working in. Nonetheless, they must ensure they are managing the risks of bribery, corruption, extortion, attempts at money laundering and whatever else goes on in the unsavoury underbelly of government/business relations that still exists in many countries where miners operate.

Cybersecurity. In a world where many mine operations are controlled remotely, multi-million-dollar contracts are executed over the internet, funds are transferred electronically, routine transactions are all online and companies just can't operate effectively if 'the system goes down', cybersecurity must be top of mind. Sophisticated hacking and attempts at hacking, either to disable systems and demand a ransom, or to steal companies' proprietary information, or just to cause chaos: these are all very real and all-too-common occurrences. Cyber threats are a major risk.

Financing, interest rates and foreign exchange. We have talked a lot about financing. Suffice to say that the risk of interest rates moving unfavourably, exchange rates moving unfavourably or financing not being accessible, or only being available on unacceptable terms, are all risks faced by miners, particularly those living on the edge of financial stability.

I could go on listing risks for pages, but I trust you get the idea. Mining is a risky business. Miners have got a lot better over the past few decades at managing and mitigating risks, but the risks will never completely go away. Indeed, most miners are working as hard as they can to minimize risk, but the only way to remove risk completely would be to stop mining. And then we would all run out of everything.

4
The economic and financial challenges of mining

There's no point developing a mine if it won't make money. Mining is expensive and risky, so investors won't part with their cash for a mine development unless there's a reasonable expectation that they will see a reasonable return. Having considered the mining cycle, we now need to consider the economics of it all.

The economics of mining are complex. Not because mining is overly complex as a concept, it isn't. As noted above, you find a deposit, you build a mine, you extract ore, you process it and then you sell it. The economics are complex because actually *doing* mining is complex. There are so many variables and the timeframes are usually very long.

From thinking about a possible mining project to getting it into production could take you ten years if there are few complexities. Run into a few hurdles and it will be twenty or more. Indeed the average development time for a large mine development is now closer to twenty years than ten. Fortunately, if it's a good project, it may have a production life ('life of mine') of fifty years or longer. The long time period before you hopefully get income from producing something means that the economics of mining projects can be fraught.

To add to all the other variables, the commodity cycle is real. Every time the pundits confidently tell us the cyclical nature of metals valuations is a thing of the past, prices plummet. So as you consider a project in 2025, hoping that you might produce some saleable metal in 2035, you really don't have a clue what the price will be. And you don't really know what the price of all the inputs will be either. All manner of non-financial factors can also change along the way, again impacting the economics.

Regulations change; environmental and community expectations change; technology and metal uses change; investor preferences change; governments and their amenability to the mining industry change. In fact, if you have a set of twenty broad financial and finance-impacting assumptions at the start of a mining project, it's reasonable to expect that they will all change before you get your mine into production.

Next time you wonder why so few mooted mining projects end up as producing mines, and why those that do take so long to get there, ponder on these. The risks are huge, the returns are uncertain and the payback period is long. When you stop and think about it, it's actually amazing that any mines get built at all. The reason they get built despite everything is that we all need what they produce. They are built because mining is essential.

I have set out below, in summary form, some of the things miners need to take into account when deciding whether a mining project has any chance of being economically viable. I make no claim that this list is exhaustive. The purpose is rather to demonstrate the extent of economic uncertainty inherent in the mining business and to help the reader understand the breadth of issues that miners, or would-be miners, need to consider. For an essential industry, the risks are huge and the uncertainties are all-encompassing. This is before you add in the difficulties of dealing with a sceptical public, overzealous activists, hypocritical investors and incompetent and inconsistent governments.

The commodity cycle

The mining industry demonstrates the immutable law of supply and demand in its purest form. When demand for a particular metal goes up, and thus there is more competition for it in the market, the price goes up. This makes mining the metal more attractive, so more supply comes on stream (eventually). As supply catches up with the increase in demand, the price goes down until the marginal supply becomes uneconomic, at which time supply falls back. If supply can't catch up quickly enough, end-users might decide the price is too high, so they either do without it, or find an alternative metal, reducing demand and thus reducing the price. Either way, the price will go up and down as supply and demand fluctuate, constantly looking for equilibrium. Commodity traders make their money by identifying fluctuations in the supply-and-demand equation before everyone else, and banking the resulting arbitrage.

The point of the above is that there will always be a commodity cycle. As demand for copper, for example, goes up, new projects will be more attractive, easier to finance and more clearly worth the effort. Those doing the risk assessments will be happier, and projects will get the go-ahead. Once these new projects come on stream, the increase in supply may mean that demand is met at a lower price, potential returns on new projects will be lower and thus fewer new projects are started. The price then finds a natural platform at which the existing producers can recoup their investment. Of course, again using copper as an example, if external factors such as the rise of electric vehicles, electrification and renewable energy mean that the demand for copper goes up beyond what might be expected due to urbanization, population increase or other historical determining factors, then the disparity between demand and supply will grow. This should mean the price goes up further, and more copper mines are built. As copper demand is often seen as a proxy for the level of economic activity, macro-economic factors, either real or imagined, such as a slow-down in growth in major markets, also impact on the copper price. The supply-and-demand-driven commodity cycle rolls on.

The tricky part with all this is that you can't just switch a copper project, or any mining project, on and off. I guess once a mine is built and in production you can 'turn it off' by putting it into care and maintenance reasonably quickly, but that isn't the same as deciding that the outlook for copper is excellent, likely to stay that way for some time and starting the ten- to twenty-year process of developing a new mine.

This brings to mind an important distinction to be made between a junior miner, or a single-commodity miner, and a large, diversified mining company. Junior miners have to consider the commodity cycle for their particular metal in isolation. Diversified miners have a bit more leeway. Diversification theory is as old as commerce itself, and it applies to mining as much as to other industries. If you are a diversified miner, and particularly if you are a diversified major mining house, the opportunity to take a calculated risk on a new development in one commodity is hedged by your portfolio of projects in both other commodities and other geographies. Making decisions on the development economics of a single mine is, by its nature, much more difficult.

As I noted above, the energy transition, the increase in renewable energy and the electrification of mobility will require more copper than current projections indicate will be available. Given that there's almost certain to be a supply shortage in the not-too-distant future, I'm often asked why the commodity price of copper isn't higher than it is. The answer is that the price of copper today reflects the demand for copper today, not the demand in ten years' time. Although there's some warehousing by producers, traders and end-users, there isn't ten years of copper stored up, so the price today represents largely the amount required for the short term. As the supply crunch gets closer, however, you can anticipate more 'hoarding', and the price will go up. Or will it?

So that's the first economic uncertainty: how to conclude, given the vagaries of the commodity cycle, that it's sensible to start on a very long-term project to extract something that may nor may not be economically justifiable when you are ready to start mining in one or two decades' time.

Processing requirements

As noted in chapter 2, processing is what you do to the mineral-bearing ore you have dug up to extract from it the metal or mineral you are seeking. The processing requirements for each type of metal are very different, and to make the processing more complicated, often ore will have more than one metal in it. Mines with multiple economically extractable metals are referred to as 'polymetallic'. In these circumstances, your processing will be complicated by the need to include the necessary steps to extract multiple metals. When one metal is dominant, the other metals which are extracted are referred to as 'by-products'. So, for example, a silver mine that also produces a small amount of gold will be a silver mine with a gold by-product. Iron ore, bauxite and coal, on the other hand, usually stand alone, or at least they are so dominant that processing specifically to extract other metals is usually not worth the cost and complication.

Processing is a necessary cost of any mining project. Even coal, which requires almost no processing, needs to be washed and sorted before being loaded onto a train. Copper ore needs to be turned into either concentrate or copper cathodes before being shipped; gold needs to be turned into either concentrate or unrefined doré bars; iron ore, unless it's very high grade straight out of the ground, needs to be concentrated (known as beneficiation); and base metals generally need to be concentrated or refined to make them economically transportable to a smelter, refinery or processing plant elsewhere. So when companies are considering a mining project, they must, from the very beginning, consider the processing requirements that will need to be developed along with the mine.

The need to process the ore you dig up therefore adds yet more complexity to be assessed, planned, constructed, operated and, of course, paid for.

Resources and reserves

At its heart, the thing you are actually doing when you mine is to extract the metal-bearing ore you are seeking from below the ground. The work of mining geologists is to determine where it is, how much of it is there and whether it can be dug up economically.

The amount of ore geologists estimate to exist within the area they are interested in is referred to as the resource. As a general rule, you might do surveys to identify an area of interest, then within the area of interest you perform exploration drilling to extract core samples. You test these samples in a laboratory and are able to determine the presence, or otherwise, of the ore you are looking for. The more drilling you perform, and the closer the drill holes are together, the greater the degree of certainty you can have about the presence and size of the orebody you are seeking. Once you have done enough drilling to make an informed view, you can say with a level of precision that there exists a resource of a certain size. How you define that depends upon the metal. Iron ore, for example, is measured in tonness. Copper gets measured in tonnes or pounds, depending upon which country you are from. Gold, not surprisingly, is measured in the much smaller increment of ounces, or, to be completely specific, troy ounces. Troy ounces are 10 per cent heavier than the normal ounces we might use to weigh ingredients for baking a cake (1 ounce is 28.3 grams; 1 troy ounce is 31.1 grams).

Once you have decided the resource exists, the next question is: how much of it can you dig up economically? It's all very well to know that sitting quietly under the ground is a few million tonnes of something, but can you actually go through all the expense of digging it up, processing it and selling it and make some money? The portion of the resource that you can economically extract is called the reserve. Reserves are thus subsets of resources.

Different countries have different standards, or 'codes', for defining and then reporting on reserves and resources. The codification of mineral reporting gives investors, financiers and others a level of certainty and comfort that the reserves and resources being reported by a company are robust and have been arrived at in accordance with a recognized protocol.

All manner of considerations must be taken into account in assessing your reserves and resources, and I will come to some of them in the next section.

The last thing I want to say about reserves and resources is in relation to their place, or rather their lack of place, in mining companies' financial reports. When I started working in this industry, it took a little time to get my head around the accounting treatment for a mining company's primary asset. You see, the reserves and resources of a mining company never appear on the balance sheet. The reason the company exists, the purpose of the whole endeavour, isn't directly attributed any value.

You might think that if, for example, you had done a lot of drilling, thoroughly examined the results, done your mine plan and concluded that you had a reserve of one million ounces of gold, at a gold price of, say, $2,000, then you could include on your balance sheet as an asset of the company two billion dollars of the yellow metal. But no, you don't. And you don't because it's impossible, with any degree of certainty, to put a realizable value on the ore under the ground, or the value that will be left once you have got it out of the ground. To use the example above, you are reasonably confident you have a million ounces of gold, but you don't yet know if you will be able to extract it all, you don't know for certain what the recovery will be (refer below), you don't know exactly what it will cost to extract it and process it, and you don't know what you will be able to sell it for when the time comes, and that time may be a long way off. So the two billion dollar figure above would be, frankly, meaningless.

What you *do* do is capitalize (that is, put onto the balance sheet as an asset) all the costs of bringing your mining project into production. You will find this on a mining company's balance sheet called something like 'mine properties' or 'mine development'. These capitalized costs are then amortized (that is, written off) against the revenue from the metal that you sell once your mine is in production. There are various other categories of capitalized costs, one of the more curious being 'deferred stripping'. This doesn't mean the mine workers are going to take their shirts off at some future time. Rather it's the capitalized cost of removing waste rock and dirt, known as overburden, from the top of the orebody you wish to access, a process known as stripping. The cost of doing this is then recouped against that ore when you eventually mine it, thus the adjective 'deferred'.

Because mining companies need to try to accurately disclose their financial situation without being able to put a direct value on their core asset, their accounting and reporting is complicated, unusual and different to enterprises in most other industries.

Grades and recoveries

Grades and recoveries are two of the significant variables facing miners looking to undertake a mining project, adding to the inherent uncertainty of embarking on a mining venture. Grade is the percentage of the metal that you actually want that's contained within the orebody that you plan to dig up. Recovery is the percentage of metal that you get out at the end of a process (be that separation, concentration, smelting or refining) compared to the 100 per cent that went in at the beginning of the process.

To use some examples: an open-pit gold mine might have a grade of 4 grams per tonne (4 parts per million); an average copper mine might have a grade of about 0.5 per cent (5 parts per 1,000, or 5,000 parts per million); and world-class iron ore deposits can be above 60 per cent iron ore (60 parts per 100, or 600,000 parts per million). As you can see, comparing grades between different metals is completely pointless. However, comparing the grade between mines of the same metal, and between deposits in mines, and even between different ends of the same deposit, is a crucial part of understanding the economics of mining and making decisions on where and when to mine.

Small changes in grade make a big difference to the economics of mining because of the amount of material that gets moved around. For example, a good-grade open-pit gold mine might produce gold at 5 grams per tonne. Put another way, that's 5 grams of gold per million, or 1 part gold for every 200,000 parts of ore. That's an excellent grade for an open-pit mine, and a well-run gold mining company should make a lot of money with an average grade of 5 grams per tonne. But if, for reasons of geology, that grade drops unexpectedly to 4 grams per tonne, still a very good grade for an open-pit gold mine, the miner will need to move 20 per cent more material, and process 20 per cent more ore through the gold plant, to get the same amount of saleable gold out the other end. And that's just the ore itself. To get to the ore, the miner may need to do a lot of stripping, removing the overburden to get to the ore they want to mine in the first place.

To put some broad-brush numbers around this: the world's biggest gold mines can each produce around one million ounces of gold per annum, a few of them even more. To minimize confusion, remember that 1 troy ounce of gold is 31.1 grams, so a million-ounce gold mine is producing 31.1 million grams, or 31.1 tonnes, of gold per annum. If we assume an average grade of 5 grams per tonne, a million-ounce mine is digging up over 6.2 million tonnes of ore from which it will separate that gold. If the grade drops to 4 grams per tonne, they will need to move an additional 1.5 million tonnes of ore. It may be hard to visualize what six or seven million tonnes of ore looks like, so let's measure it in trucks. We aren't talking normal-size lorries. A haul truck in an open-pit mine probably has a capacity of at least 200 tonnes and is the size of a three-storey building. So over six million tonnes of ore to be transported from the mine to the processing plant requires approximately 31,000 of these enormous loads. At 4 grams per tonne instead of 5, you would need to process an additional 7,500 haul trucks full of ore just to sell the same amount of gold. And remember, too, that this is all before stripping. If the orebody is lying under 50 metres of non-gold-bearing rock, all that rock needs to be removed ('stripped') and sent off to a waste dump before you can start mining the gold ore. That's a lot more truckloads.

There are profitable gold mines in some parts of the world which extract gold ore at less than 1 gram per tonne. I have no idea how they do it, but they do. They are moving a lot of dirt to make their money!

Recoveries have a similar effect. If your gold processing plant is producing gold at 95 per cent recovery, that means that the process is extracting 95 per cent of the gold that was contained within the ore that went into the crusher at the start of the process. The other 5 per cent ends up in the tailings, or waste. It's very difficult to produce gold without some measure of precious metal remaining in the tailings, which is why a lot of mines build tailings retreatment plants, which reprocess the tailings, seeking to extract economic quantities of the gold that was left in the waste the first time around.

If the metallurgy of the ore, or the effectiveness of the chemicals used to treat it, or some other factor changes, then your recoveries may be impacted. If the recovery drops, for example, by 10 per cent, you will be producing 10 per cent less gold for exactly the same effort, and thus, all other things being equal, the cost to produce an ounce of gold will have gone up by 10 per cent without you doing anything.

Changes in grades and recoveries impact the economics and thus the profitability of your mine. Putting it all together, figuring out whether the reserves and resources you have identified, at the grades and recoveries you expect to achieve, are really going to underpin a profitable mining project is a very complex business and is critical for deciding whether to proceed, how to proceed and, very fundamentally, successfully obtaining funding for your enterprise.

The remoteness factor

When you visit Johannesburg in South Africa, you are struck by the remains of the waste dumps from the gold mines which were the original reason for the city's establishment back in the 1880s. These hills, identifiable as waste dumps because of their unnaturally rectangular shapes, are the products of many years of very profitable gold mining in what is now the centre of a very big city. They are also notable because it's so unusual to see the remnants of mining so close to a major urban centre, although, of course, the mines were there long before Johannesburg was a metropolis.

And so to another of the causes of economic uncertainty in mining. Mines tend to be in remote places. Not always, but usually.

The world's richest copper region is in the Atacama Desert in the north of Chile, also as it happens the world's driest place. African copper comes from places like the Democratic Republic of Congo, not exactly an hospitable part of the globe. Gold comes from all corners of the world, including the sparse, frozen reaches of Russia, the baking deserts of Australia, the almost inaccessible jungles of Papua New Guinea and the badlands of West Africa. Australia's iron ore comes almost entirely from the Pilbara, about 1,000 miles north of Perth, the nearest major city and itself not exactly close to much else. I once visited the Tanami Gold Mine in Australia's Northern Territory. Getting there involved flying to Alice Springs, a town in the middle of absolutely nowhere in the very centre of Australia, and then taking a light aeroplane (a quite distinctive aircraft known as the 'flying pencil') 600 miles into the Tanami Desert.

I still remember one of the more unusual flight safety procedures that was a routine part of landing at the mine's airstrip. After an uneventful flight north-west from Alice Springs, we descended towards the runway. However we didn't immediately land. As is common with bush airstrips, the pilots executed a low pass so they could do a visual inspection of the strip before landing on it. What was perhaps unusual about this procedure was that the low pass had the added and quite vital role of scaring any grazing cattle off the runway before we landed. I'm not entirely sure what would have happened if one of the cattle had absent-mindedly decided to return for some more succulent runway grass in the time it took the plane to 'go around' and come back in for the actual landing.

Until I started flying to the Russian Far East, I genuinely thought the Tanami was about as remote as it was possible to get. The wilderness of northern Canada, the uncertain expanses of much of Central Africa, the steaming jungles of Brazil, the remote steppes of Mongolia, the inhospitable mountains of Chile and Argentina: mining seems to go out of its way to be hard to get to.

Which brings me nicely to another flight-related experience, this time in the seriously remote and frozen wilds of the Russian Far East I referred to above. After flying eight hours east from Moscow to Blagoveshchensk, my team and I boarded a helicopter on a very snowy day for what should have been a two-and-a-half-hour flight north-east into the mountains to visit yet another gold mine. About an hour into the flight, the snowfall became a blizzard and the visibility became too low for the pilots to safely continue. Fortunately, we were quite close to the mining company's warehouse and staging post at the bottom of the mountains and were able to land there before the conditions deteriorated completely. A very rickety old minibus at the warehouse was duly commandeered and we set off by road through the thick forest. The landscape was incredibly beautiful, the tall dark pines the only things that weren't pristine white. White snow-encrusted road, white snow piled high on either side, white covering every non-vertical surface of the trees, and white sky with thick snow falling heavily and swirling about in our wake. The growl of the minibus's engine was the only disturbance in what would otherwise have made a perfect real-life snow globe.

At one point, we reached an icy river and crossed it on a cable barge that was even more rickety than the minibus. Made of timber, and with an ancient diesel motor to haul itself across the river, it shuddered and belched and clanked and creaked but nonetheless did the job. Despite being at least minus 25 degrees outside, we all had to alight the minibus and stand on the deck of the barge. This was for the very practical reason, it was explained to us, that it was safer to be on the deck rather than on the bus in the event that the barge should start to sink. Very comforting. Eventually, we made it to the other side, re-boarded the bus and continued on our way through the seemingly never-ending pine forest.

The snow was no longer falling so heavily and it was once again safe to fly. Now came our spy-thriller moment. As we trundled along the road through the forest in our arthritic minibus, suddenly as if from nowhere appeared the helicopter, descending slowly and noisily to land in the middle of the road in front of us, its rotors sending the fresh white snow flurrying absolutely everywhere, their staccato racket amplified as it ricocheted off the mountainsides.

Once it was on the ground, we were quickly bundled out of the bus, into the helicopter, and within ten seconds of the last person scrambling aboard, the door was slammed shut and we were airborne. Amazing. And also quite a relief. The thought of being stuck in probably the remotest place I have ever been (the Tanami Desert included) in the sub-arctic cold in a broken-down minibus wasn't a pleasant one.

I now realize that just when you think you can't get any farther away from everything, somewhere even more remote will beckon, probably with a mine in it.

The remoteness of much of the world's mining adds all manner of economic considerations. Assuming for a moment that someone has concluded that there's at least the potential to economically extract minerals from a location, you then have to think about the following questions. Is the location accessible, and not just in a helicopter? Would you be able to get the materials and equipment in there to build the mine, any necessary processing plant and accommodation for workers? Is there infrastructure nearby that can be accessed or connected, such as roads, water, power, railway lines and ports? Does it rain too much or too little? Too much and you will be constantly dealing with flooding and removing water from pits (known as 'dewatering'), too little and you will need to factor in very expensive pipelines, perhaps desalination plants and other infrastructure. Is the area seismically stable – not known for earthquakes? Is it constantly shrouded in fog so you could never land an aeroplane or a helicopter? Is it so high above sea level that you would be constantly dealing with real or potential altitude sickness?

I once visited a copper mine up in the Andes in Chile. We flew in a small aircraft to a landing strip at 2,500 metres above sea level and drove to the mine office at 3,000 metres. Before we were allowed on site, and up to the top of the mine at 4,000 metres, we were checked over by a doctor at the mine clinic to ensure we weren't an overt altitude sickness risk. So one also has to consider questions such as the following. What's the healthcare availability? Will you need to build your own hospital? Is there a ready workforce, or will you need to fly everyone in from somewhere else? 'Fly-in fly-out' can work – it's pretty standard in the Pilbara – but it's expensive, is a significant logistical exercise and comes with a unique set of social and domestic challenges.

There's a significant economic and organizational cost to developing and operating a mine in a remote location.

Construction costs

Before you can produce a single tonne of iron ore, a pound of copper or an ounce of gold, you need to build your mine and everything that supports it. How much that will cost you is down to a dizzying number of variables, including the remoteness or otherwise, as discussed in the preceding paragraphs. I mentioned in chapter 2 many of the elements of a mining project that need to be built. In summary, there will be the mine itself, then all the crushing and processing plant, the infrastructure, accommodation, administration blocks and, potentially, community assets. Mining requires a lot of construction.

And the construction doesn't come cheaply. A few recent examples: a recently constructed iron ore mine in the Pilbara cost $5 billion; a new copper mine recently completed in Peru cost $8 billion; a massive greenfield copper mine in Russia which is in the early stages of construction is also anticipated to cost about $8 billion; and a recent iron ore mine in Brazil ended up costing about $10 billion. Even a small gold mine will set you back a few hundred million dollars. And all this money needs to be found, self-evidently, before any metal is produced and thus before any revenue is earned.

Ironically (pardon the pun), the cost of constructing an iron ore mine and all the associated plant and transport infrastructure will be materially impacted by the price of steel, the very product the iron ore you wish to mine is destined to become. So the better the prospects for your iron ore, the more expensive your mine will be to construct in the first place. Sometimes you just can't win.

A subset of construction is the work of either connecting to or developing standalone infrastructure. For example, the mine will require power, either through connection to a grid, or by building a generation plant. The mine will need to connect to a water supply or access its own. It will need to connect to a road network, which, depending upon remoteness, may involve building a road just out the gate, or may extend for hundreds of miles. It may need to build an airstrip. If it's producing bulk ore such as iron ore, bauxite or coal, it will need to connect to the rail network or perhaps build its own railway to the nearest port. It may need to build a pipeline, for example the 500-kilometre slurry pipeline to ship iron ore from Anglo American's Minas-Rio mine to the port. It may even need to build the port!

Suffice to say at this point that when considering the economics of a mining project, the costs and complexities of construction will be a huge factor in determining whether to proceed. Just as importantly, a plausible story on construction costs will be essential before investors will part with their money, and will be particularly essential before convincing lenders to part with theirs!

Community contributions

When mining companies are considering projects, particularly in developing countries, there's often the expectation, or indeed a licensing requirement, for the company to contribute to the community in which the mine will sit. This is in recognition not only of the inconvenience the mine will cause, particularly during construction, but also of the fact that as the mining company is extracting valuable metal from the ground, it should 'give back' to the community which has historically lived and worked on that ground. This is often referred to as 'the social licence to operate'. Of course, it's also important to note that a mining project will usually make a natural contribution to the local community by providing employment together with the 'multiplier effect' of those newly employed workers then spending their money in the community.

These community contributions come in a variety of forms but common ones include: the provision of power; the provision of a reliable, clean water supply; constructing roads or other infrastructure; building hospitals, clinics, schools or other public buildings; and providing services for these new facilities, such as medical staff and teachers. Sometimes the contribution will also include replacing facilities that have been lost because a mine is being built on top of them. Not all mines require the diversion of rivers or the literal removal of mountains, but it happens, and when it does, the company will need to sort all that out and replace the roads, bridges or indeed whole villages that have been displaced.

These community contributions always work best when there's plenty of ground-level consultation. I remember hearing of one classic example of what can happen when there's insufficient input from the community a miner is trying to help. At the opening of a new school, proudly paid for by a London-listed mining company constructing a new mine in Africa, the local community leaders noted that they were very grateful for the new school but what they would actually have preferred, and what would have been cheaper for the mining company, was simply a decent bridge over the adjacent river. That way their children could have gone to the very good, long-established school on the other side.

It's reasonable to say that miners have become a lot better at community engagement and the vast majority now treat it with the seriousness it deserves. Indeed, like the mine itself and the related infrastructure, community engagement and the development of community infrastructure is now dealt with as a project in its own right.

Labour costs

Up until this point, we have been largely focused on the economics of building a mine. With labour costs, we come to the key costs of running a mine. Mining continues to be a labour-intensive industry. Not nearly as labour-intensive as it was before the age of mechanization, but it requires a lot of manpower nonetheless. Labour costs usually make up between 20 and 30 per cent of a mining project's total ongoing cash costs.

The cost is often exacerbated, once again, by the remoteness factor. On the positive side, remote mines often provide stable, valuable employment for local communities for which there may otherwise be few options. Much has been made over the years about mining companies allegedly exploiting local communities, and there have unfortunately been some pretty bad examples of this. But on the whole, the presence of a mine usually brings opportunities for gainful employment which, agriculture aside, wouldn't otherwise exist.

Where a local workforce isn't available, or isn't sufficient, a mine will need to 'import' its workers. Either driving them in or out, or flying them in or out. Perhaps one of the most extensive 'fly-in fly-out' operations is the giant Pilbara iron ore mining region in Western Australia. I have often sat in the Qantas Frequent Flyer lounge at Perth Airport and been one of the few people in a suit, surrounded instead by people in hi-vis work gear on their way to the Pilbara on fly-in fly-out charter flights. I have also visited a gold mine in Russia which, on changeover days, shuttles its workers by helicopter down the mountain from the mine to the nearest railhead.

All this costs money. Not only do you need to fly them in and out, you also need to house them. Accommodating miners these days means mine camps with private rooms, good facilities and, as I have been privileged to experience, seemingly limitless quantities of fabulous food. Miners work hard and they eat like there's no tomorrow.

The final issue with labour costs is market forces. Attracting mine workers, particularly to remote locations, will often require above-average wages. Indeed, during the boom years for mining in the Pilbara, such was the demand for workers that the guys who swivelled the 'Stop/Go' signs directing traffic at mine haul road intersections were earning over double the salary of a junior doctor back in Perth. This had some unintended consequences. During these boom years, I had a client that owned and operated sugar mills along the tropical coast of Queensland. They had great difficulty attracting qualified tradespeople to work in the sugar mills – electricians, fitters, mechanics, those sort of things – because they simply couldn't match the wages offered by the mining companies in the Bowen Basin coalfields, a few hours' drive inland.

The labour requirements for mining are also changing. As mines embrace the digital age, they need employees with digital skills, and this means competing with the IT and broader technology industries, again potentially increasing competition for talent, and thus salaries.

Contracting costs

Many miners don't use their own workforces for all the elements of their mining operations. For example, a lot of mines have the actual mining, as in the primary digging up, done by contract miners. Other mines use contractors to transport ore, or to maintain the vehicle fleet, or provide security, or maintain plant, or even run the plant. Just about all miners use contractors for blasting rock. This is a good thing. Outsourcing your blasting to a company that spends its day thinking about nothing else but the safe preparation and execution of massive explosions makes a lot of sense. Once you add all these functions together, the majority of the workforce in many mines are contractors.

There are pros and cons with contracting out the various functions involved in the mining process. You gain the benefit of using contractors with specific areas of expertise. You gain some financial certainty as you have established contracts with either known costs or at least known cost mechanisms. You may, depending upon the contract, gain flexibility in that you can quickly turn the supply of whatever you have contracted on or off. But you may also lose an element of control, and you may lose some efficiency, and you may lose in-house expertise. You may also lose some profitability as the contract will (hopefully for them) include some profit margin for the contractor. It also becomes more difficult to ensure the whole mine workforce adheres to the company's operating and safety standards because a chunk of them don't work directly for you.

So the determination of whether to owner-operate or contract out is another variable to be considered in running your mine.

Energy costs

It's slightly ironic that mining, which is an essential part of the transition to a more sustainable energy future, is itself such an energy-intensive industry. Hard-rock mining requires a lot of power of different sorts. Removing thousands, often millions, of tonnes of rock and dirt to get at the ore requires the power of excavators run on either diesel or electricity. Then you need a whole battalion of haul trucks to move material around: waste to the waste dumps, ore to the processing plant or to stockpiles. These run on diesel, although there's a lot of work going into alternative forms of power, including electricity (using catenaries – overhead wires – like a train, or batteries) and hydrogen. There's electricity required to run the crushers, the mills, the processing plants and the sundry other moving parts inherent in a mining operation. Then there are the furnaces to smelt things, run on coking coal, gas, oil or electricity. And finally there will be the electricity required to run everything else: plant, lighting, ventilation, accommodation, offices, the IT systems, the water systems and much else. All this electricity will come either from a public grid or, in more remote areas, from an on-site generation plant, probably run by coal, oil or gas but these days also possibly from renewable sources such as solar or wind power or, more often than you might think, hydro generation.

And all that power costs money. As it's a significant proportion of the running costs of a mine, roughly 30 per cent, crucial assumptions need to made about how much will be required and how it will be sourced, as part of any viability assessment.

Making this assessment is made more difficult at the moment because of the uncertainty around future energy supplies. For example, if the mine is in a western country, will it still have access to reticulated gas in ten years' time, or will it have been banned? What will be the price of oil in ten years' time, particularly if there's government intervention by way of carbon taxes, non-market limitations on supply or punitive regulation? What will be the price of national network electricity after the electrification of transportation, heating and everything else has increased demand without a consequent increase in stable supply? What assumptions can be made about the cost of renewables? Will the cost come down as technology develops and stability of supply improves? And to what extent can we incorporate into our planning ongoing energy efficiency gains in a mine's operations? Forecasting energy requirements and costs ten or twenty years into the future is difficult, perhaps almost impossible, to do with precision.

So you can add energy costs as one of the great uncertainties afflicting the accurate assessment of the viability or otherwise of potential mining projects.

Equipment and consumables

Mining operations use a lot of 'stuff', to use a technical term. Any operating mine will have a whole array of equipment, including vehicles of all sorts, from the enormous haul trucks, industrial-size diggers, 'jumbo' drills, water and fuel tankers, all manner of trucks for different purposes, down to domestic-size pick-up trucks. They will have drilling equipment, ventilation equipment, electrical reticulation equipment, maintenance equipment, safety equipment, refuelling equipment, portable measurement and analysis equipment, catering facilities, all sorts of personal equipment for the workforce, transport equipment – you name it, a mine will have it. And all this 'stuff' needs to be maintained, occasionally refurbished, even more occasionally replaced, and regularly filled with fuel and lubricant and refitted with new tyres, brake pads, bushes, filters and so on.

Depending upon the metal you are mining and the consequent processing requirements, you may also consume a dizzying array of chemicals. Copper mining, for example, uses an impressive amount of acid. A young accountant who used to work for me left my old firm at a relatively early stage, moved to Switzerland and joined a commodity trading company. He started off not on the coal, iron ore or crude oil trading desks, but on the desk trading sulphuric acid. He made a lot of money. Now I know why. The demand for acid in the mining industry is big. Occasionally, one hears of restrictions in processing caused by periodic shortages of sulphuric acid. Gold mining, on the other hand, uses a lot of cyanide in the process to separate the gold from the host ore. The gold industry's reliance upon cyanide troubles some people, but in responsible hands it's actually very safe. Of bigger concern from a toxicity perspective is probably arsenic, which is often present in rocks that also contain precious metals, and so forms a common waste product of the gold mining process that ends up in tailings dams.

So a mining operation needs equipment, consumables and chemicals. And miners need to factor all this into their plans, at least five years before they will start to use any of it.

Water

Mines have traditionally used a lot of water. Most mines still do. However, one of the unheralded technological advances of the mining industry in recent years has been the development of mining processes which require far less water. This is another example of the mining industry's quiet contribution to sustainability. Often, it has to be said, the drive to reduce water consumption is not only desirable but also a practical necessity. Mines are often in arid regions. For example, the Atacama Desert in Chile, home to many of the world's biggest copper mines, including Escondida, the biggest, is also, as noted above, the driest place on earth. So water is a particularly important issue. Miners have addressed this in two ways: developing processing that requires far less water; and using seawater, pumped hundreds of miles from the ocean, which they either have adapted their processes to use directly, or put through a desalination plant.

The point is: water is an essential and scarce resource. The cost of obtaining it will probably continue to increase. The alternative cost of developing new technologies so you don't need as much is the more sustainable option but it's still a cost. Either way, mines will always need water, if for nothing else, for the workers to drink.

Environmental costs

As we survey history, it's valid to blame sections of the mining industry for having a cavalier approach to the environment. There's a sad litany of both intentional and unintentional environmental disasters, from the pollution of rivers (which were often also the local communities' water supplies) to the pollution of the atmosphere, the destruction of forests, the removal of habitats and even, tragically, catastrophes such as tailings dam collapses.

These days, the vast majority of miners are environmentally responsible, and indeed go out of their way to ensure that not only are they responsible, but also they actively manage risk to minimize the danger of something going wrong. The difficulty for miners is that, as with everything else they do, no-one notices when things go right, but everyone notices when things go wrong. On the upside, this is certainly further incentive, if any is needed, to focus on the very best environmental outcomes.

Any mining project these days will require an environmental impact study (EIS) as part of the fairly early stage of planning. This will need to be approved by the relevant regulator, and is quite an exhaustive exercise which takes both time and money. An EIS isn't just to tick a regulatory box. It's an opportunity for those seeking to develop a mine to ensure they will be doing so in a sustainable way. But that doesn't alter the fact that for a big project you probably need to allow at least eighteen months (and probably a lot longer) and a fair bit of cash to get it done. And that's just at the planning stage.

Ensuring you have ongoing acceptable environmental outcomes takes planning, care, time and, of course, money. Investing in renewable energy; implementing process improvements to reduce water usage; adopting measures to reduce emissions and generally seeking to reduce a mine's 'footprint': these are all important things to do, but they all cost a lot of money. These need to be factored in to any mine development, with a pretty big contingency to allow for future environmental regulation, investor expectations and societal norms. You then have the ongoing operational cost of monitoring, managing, reporting on and improving your environmental impact.

Unfortunately, what is regarded as an acceptable environmental outcome differs depending upon your view of the world, your acceptance of the need for trade-offs, as discussed earlier in the book, and probably your politics. If you accept that mining is essential, then you will cope with a certain amount of environmental disturbance, as long as it's contained, there are no damaging side-effects (for example air or water pollution) that spread beyond the immediate area of the mine and it can, at the end of the life of mine, be properly rehabilitated. If, however, you are an environmental absolutist, none of that will be acceptable to you.

Technology costs

In improving the sustainability of mining, countering the impact of declining grades through improved productivity and efficiency and simply making more money for investors by employing the best, most up-to-date mining and processing techniques, miners are busy improving what they do through technology. Technological innovation is a big driver towards more sustainable mining, and I cover it in more detail in the next chapter.

Technological advance is an important factor when considering the economics of mining. Miners need to constantly invest in innovation, develop and trial new technologies and experiment with new mining and processing techniques. And they need to spend all this time, effort and money well in advance of actually getting a return on their investment.

Management, reporting and indirect costs

You will have realized by now that there are an awful lot of facets to the mining business, and developing and then running a mine is complex and intense and involves numerous activities going on at once. It therefore needs a great deal of managing. We referenced the importance of good management in the previous chapter. So as a mining project grows from a venture of perhaps half a dozen people looking at a geological survey and thinking they may have found a magnificent deposit, to a newly developed operating mine with a team of hundreds or thousands of employees, they will need to think about all the people needed in addition to the actual miners themselves. This will include, of course, people to run the company, starting with a board, a CEO, a Chief Financial Officer and the rest of management.

To enable that managing to happen, a mine not only needs managers, it also needs support functions like finance, IT, human resources and all the other internal teams that make sure everything happens as and when it should. It will also need external-facing teams like procurement and, if it's a public company, investor and public relations teams to make sure everyone knows what a great job the company are doing. This means those half dozen people I mention above may wake up one morning and realize they now employ thousands of people without really noticing how it happened.

It's not just people. A mining operation also needs IT systems for management, process control, financial management and reporting, trading, maintenance and fleet management, human resources, legal, security and everything else. And it needs offices, warehouses, other buildings, vehicles, maybe aircraft, and all the bits and pieces that go with being a mature operating business.

This needs to be planned for and then all these people will want to be paid. So add a payroll system to everything else, by the way.

Never let anyone tell you mining is simple.

Interest and other financing costs

When I was planning this chapter, base interest rates, or the 'risk-free rate', was close to zero in many countries. Now it isn't. But regardless of the base rate, it's a truism of finance that the more you need money, the greater the margin the financier will want and thus the higher the interest you will pay. So those who are low risk and have lots of options can get money very cheaply. Junior miners, on the other hand, who need money more than most, will need to factor in a rather higher financing cost.

As I discuss in more detail below when considering financing, there are in summary three ways to finance a mining project. You can either raise equity, obtain traditional debt financing or arrange a lump sum upfront with which to pay for your development, in return for making production or profit-based payments once your project is commissioned and generating offtake (production, usually used in the context of 'offtake agreements', where you commit to sell future production to a specific purchaser at an agreed price).

So when determining the economics of a mining project, anything other than straight equity will require the company to take into account the financing cost. This will comprise interest to be paid through the years when there's no income, and probably interest to be paid for some time when there is income, or income that will be forgone once the cash starts to flow, or a combination of both. It will also include some pretty hefty fees paid to bankers and advisers for the privilege. However a project is financed, a miner will usually need to factor in a significant finance cost, along with everything else, when figuring out if their proposed development makes economic sense.

The key thing to remember here is nothing is free, particularly not money.

Marketing and trading

The beauty of commodities, by their nature, is that there's always a market for them, even though that market may wax and wane. The waxing and waning is why commodity prices go up and down the way they do. Having an established market doesn't mean, however, that there aren't costs involved in selling them. Many miners have established marketing hubs, often in Singapore. These marketing operations are the means by which producers maximize the price they get for each tonne they ship. The marketers will negotiate volumes, pricing and terms; agree delivery dates and quality parameters (for example 65 per cent iron ore with a certain moisture content); and determine shipping arrangements. They will then look to optimize the sale by, for example, ensuring they make the most efficient use of bulk carriers going in the same direction, find return cargos if possible, so they don't sail back the whole way empty, and so on.

The next logical step for the marketing teams has been to use the expertise and information flows they developed initially to sell their own product to do the same with third-party product. And thus many of these marketing hubs have turned into trading hubs. There's a lot of money to be made in trading, particularly when there's volatility in commodity prices, but there's also a heightened level of risk to be managed.

A key role of marketing and trading departments, and in fact so key that often this will be a board policy matter, is hedging, or not hedging, as the case may be.

When you are producing output, and you don't already have offtake agreements in place which may already have price arrangements in them, it may be that you decide to 'hedge' against the risk of the spot price going down. So you agree, usually with a bank, to sell a certain number of ounces, at an agreed price, for delivery at an agreed date in the future. As you have a certain level of costs, you want to make sure these will be covered in the event, however unlikely it may seem, of the price for your commodity falling significantly. This may seem like a sensible risk-mitigation strategy, and it is. However, there are a number of things you need to consider.

Firstly, hedging carries an opportunity cost. If you hedge at a certain price, and then the price goes up, not down, you will still have to sell the agreed amount at the hedge price, so you will miss out on the benefit of the rise in price.

Secondly, you need to consider how much you should hedge. And this will depend upon the precariousness or otherwise of your cash position. If you are a diversified miner with plenty of cash in the bank and a portfolio of assets, perhaps in a range of commodities, then you probably don't need to hedge. You can absorb the loss of cash resulting from the drop in price for one commodity for a period of time. Being unhedged also gives you more flexibility. For example, if you are producing gold, as we talked about above, then you might decide to slow down production until the price goes back up, or keep on producing (assuming you have the cash) but stockpile the gold for a while rather than selling it. And, of course, by doing this you remove the opportunity cost I mentioned above.

But if you are a single-commodity producer, and you don't have a lot of cash to carry you through lean times, or you are heavily indebted because you borrowed heavily to build your mine and processing plant in the first place, then you need certainty of cashflow. So you may decide to pursue a hedging strategy. Then the question becomes how much, or what proportion of your planned production, should you hedge? Hedge too little and it may be ineffective. Hedge too much, however, and you may run the risk of not being able to deliver against the contract, for example if you run into a production problem at your mine.

The marketing and trading function isn't really an add-on, it's a key part of ensuring you make the right decisions on how and when to sell your production, optimize your returns, protect yourself against unexpected price fluctuations and perhaps, at the same time, trade on the swings in the market.

Mine closure and rehabilitation

Even the greatest of mines will eventually run out of ore. And then you need to close them. The Quellaveco copper mine in Peru has recently commenced production and potentially has a 100-year life of mine. That's fantastic. But even so, in about 100 years from now it will close. When a mine closes, there's a lot you need to do. Firstly, you need to remove the processing plant, tanks, offices, pipelines, conveyors, crushers and so on. Some or all of the buildings may be able to be repurposed for some other activity, perhaps manufacturing or the processing of third-party ore from somewhere else, but otherwise most of them will need to go.

Then you need to make the mine safe, both physically (so no-one can fall into a pit or down a mineshaft) and environmentally. If there are tailings dams with waste that includes dangerous chemicals like cyanide or concentrated arsenic, you will need to be satisfied that they won't leak and leach poison into the ground and potentially into the groundwater.

Finally, you will need to rehabilitate the site: planting trees or other vegetation on the waste dumps; perhaps letting the open pits fill with rainwater to make a lake, although you need to make sure the lake won't then be poisonous from all the blasting residue and other chemicals; and generally seeking to return the site to whatever its natural state was before the whole thing started. Sometimes the relevant government environmental agency can get a bit carried away, and demand that mine sites be, in effect, over-rehabilitated. There's a standing joke that in the Australian outback you can tell where the old mines were as you fly across the consistently arid, scrubby, reddish-brown landscape. The former mine sites are the only green bits with trees.

All of the above costs money, and to make sure you have this money at the end of the life of mine, the closure and rehabilitation costs need to be factored in right from the planning stage. The 'closure provision' is a line in the accounts of a mine for this reason. In many countries, miners are required by legislation to lodge a bond, or bank guarantee, to ensure there are sufficient funds available for this purpose.

The last thing to mention, but the most important, is consideration of the impact on the community of a mine closure, and how that should be planned for. What will all the locally employed mine workers do once the mine has closed? What will the providers of goods and services do once the mining stops and the mine doesn't need their food and other provisions, their delivery drivers, their office cleaners or their tyre-fitters? And then what happens to their families who relied upon them? And who will pay to maintain the public services, such as the hospital and the school, which the government happily abdicated responsibility for all those years ago when the mine opened?

There's a lot to consider. Are there alternative sources of employment that can be developed, such as agriculture? I know of a mine in Sierra Leone that went on to become a giant pineapple farm. Cattle grazing is another common use for former mine sites. The point is, the question of the post-life of mine of the local community is an important consideration for any mining development and it needs to be taken into account, along with everything else, from the planning stage onwards.

~~~~~~~~~~

We have noted previously that in the mining game the need to spend a lot of cash starts a long time before any revenue starts to flow. So financing mining can be a serious challenge.
~~~~~~~~~~

The first thing to consider is the right mix of funding.

There are many ways of funding a mining project, but they boil down to three broad types: equity, either at the company level or the individual asset level; traditional debt from a bank or other financial institution; or a more participatory arrangement where an institution will provide a lump sum upfront with which to pay for some or all of your development, in return for receiving production, royalty or profit-based payments once your project is commissioned and generating offtake.

The simplest way to raise funds is to sell equity. If you are a single mine company, then this is straightforward. If you have more than one mine, then you may wish to sell shares in the group, or you may wish to sell a portion of the shares in a specific mine. Even the big diversified mining companies do this regularly. As a way of spreading the cost of development and risk, many, indeed probably most, of the big mining projects have minority shareholders. For example, most of the big copper projects in Chile have Japanese trading houses holding non-operating stakes of between 20 and 40 per cent. From the minority shareholder's perspective, these stakes are valuable because it gives a seat at the table and the ability to guarantee offtake, for example for their trading operations or to feed downstream processing in which they may also have an interest.

Even where a mining development proposition is attractive and equity would be readily obtainable, it may be that the existing shareholders don't wish to be unduly diluted and so would prefer debt funding, particularly if it can be obtained at low rates. So equity isn't always the answer. Dilution through raising equity means giving away both future dividend streams and capital growth. An additional factor to consider is that if the equity you have sold ends up concentrated in the hands of a small number of shareholders, it may threaten your control of the company in the future.

But sometimes this is a deliberate strategy. For example, if you are an explorer, you have found what you believe to be a fabulous deposit but you have no cash to prove it up, you may be very happy to offer an 'earn in' to an investor willing to pay for the necessary bankable feasibility study in return for a significant chunk of equity. If the feasibility study confirms that you have a deposit worth developing into a mine, then the increase in value of your remaining stake in the project will be far greater than the equity you gave up to fund the study in the first place.

Debt funding, either from a bank, some other financial institution or through corporate bond raising, is the second traditional form of financing. As we have noted previously, a lot of junior miners struggle to raise debt in this way because of the understandable caution of many financial institutions to what is a high-risk, delayed-return venture. Debt funding has the advantage of not diluting existing shareholders; it has the big disadvantage, however, of requiring regular interest payments starting long before any revenue flow commences. And for junior explorers and miners, this may be simply not manageable, even if they could find a bank willing to offer it.

Given the difficulty, particularly at the junior end, of obtaining traditional finance, miners will often raise funds through other arrangements, all of which have the common characteristic of cash being provided upfront in return for giving away a portion of the value the mine ultimately produces. This could be in the form of an agreed dividend stream; a percentage of the actual offtake; or a percentage of the profit, future equity or some other agreed portion of the mine's future value, either physical or financial. The institutions providing these forms of financing often do it because they are looking to lock in guaranteed offtake, for trading or their own processing requirements.

The downside of streaming and other 'cash now, give away value later' arrangements is that if translated into an effective interest rate on your initial borrowings, it can look pretty extreme. And if the agreement is for a royalty stream for the life of the mine, then it's like having a loan that you never finish repaying. Of course, if the alternative is that you don't get to build the mine at all, this may seem like a reasonable, or 'least worst', option.

So whatever form of financing is adopted, there are difficult trade-offs to be made. Do you give up some control, and give up future dividends, for equity? Do you go through the trauma of seeking bank financing or the convoluted exercise of a corporate bond? Do you give away future value for an interest-free lump sum now? I guess it depends upon your risk appetite, your banker's risk appetite, your expectations of future profitability, your desire for control, the timeframe and your long-term plans for the company.

Financing mine developments is difficult, made more difficult by the length of time before they have any revenue. It's yet another of the impediments to getting mining projects over the line. Given the increasing demand for metals to, among other things, feed the unstoppable growth in renewable energy, I hope the providers of mining finance will do their bit for sustainability by keeping the funds flowing.

~~~~~~~~~~

It's not just the mix of funding which is a challenge for mining companies. If you are an established miner, listed on a recognized exchange, it's likely that a significant percentage of your share register will be taken up with institutional investors, and that brings with it its own challenges.

One of the most significant changes in the institutional investor landscape of recent years has been the rise of activist investing even from established funds. It's now common for major global funds to express their position on various 'ethical' matters such as holding investments in thermal coal, oil and gas, tobacco, defence industries or even junk food. I had a conversation recently with a representative of a very large New York-based family office with billions of dollars under management. They told me they were very happy to consider investing in what they cheerfully termed 'sin assets'. After all the confected virtue
~~~~~~~~~~

signalling by so many of the mainstream funds, I found this approach strangely refreshing.

The best outcome for the sustainability agenda is for those assets which are regarded as the most problematic to be the most public in their ownership and governance. In the mining industry, this is particularly relevant to thermal coal. Even as the transition to renewable energy gathers pace, the demand for thermal coal won't go away overnight or indeed probably for decades. So we should be encouraging the continued transparent ownership of these assets to maximize accountability around safety, sustainability, environmental performance, closure planning and rehabilitation. Where listed companies, for strategic reasons, wish to divest themselves of such assets, they should ensure they are either demerged into equally transparent and accountable listed companies or sold to new owners with similar standards. Selling thermal coal assets into unaccountable private ownership or into the opaque ownership of sovereign wealth funds will simply end all accountability and any expectation of a more sustainable outcome. The unseemly 'rush for the exit' currently advocated by many funds may end up having exactly the opposite effect to that which they claim to be seeking.

Of course, mining has it easy, in some respects, compared to oil and gas. The world will need oil for a long time to come, so why investment funds are rushing to demand the end of oil extraction by listed companies is beyond me. You can either have the oil industry in the hands of well-regarded, publicly reporting, responsible oil and gas companies, or you can force it into the realm of unaccountable, opaque and often not particularly ethically focused private companies. Or, even worse, you can leave it in the hands of the national oil companies of countries whose regard for transparency is on a par with their regard for the environment, for human rights, for full disclosure and for anything other than propping up their often oil-dependent economies to ensure the flow of corruption-tainted income keeps flowing.

Another issue investment funds should reflect upon is the risk of betting on the wrong technology. A lot of current technology will soon be out of date as innovation in the 'tech' space moves on very quickly. I would encourage investment funds to use some of their new-found activist energy to encourage research into innovation in the field of sustainable energy.

The good news for mining companies is that, whatever the long-term solutions for more sustainable energy turn out to be, there will be plenty of metal involved.

Returning to the implications for listed mining companies, the question then becomes: how does a listed mining company ensure they can proceed through their corporate life with minimal drama from, particularly, institutional shareholders?

Firstly, be clear about what you are trying to achieve. If activist funds are questioning your strategy, explain it and be robust in your defence. For reasons that I don't really understand, most of the noise around the demand for divestments has been from one direction. The argument for keeping problematic assets in responsible hands has been made relatively quietly by comparison. Sadly, then, the outcome is that the demands being made by some of these funds are unrealistic and unhelpful. So stand up to them.

Secondly, ensure your own directors, management and staff understand your strategy. And make sure that your own team understand the essential nature of mining. You could get them all to read the first chapter of this book, for example. But seriously, a lot of 'corporate wokeism' is being driven by CEOs' fears about upsetting their own staff. If your people understand that a hasty divestment strategy will almost certainly have bad sustainability outcomes, their support for a more measured approach will be much easier to obtain.

Thirdly, ensure your company aren't themselves being hypocritical, by ensuring they are maintaining high standards in safety, sustainability and transparency themselves.

It's important to note that the concerns I have raised in the preceding paragraphs about activist investors seeking to deny equity funding to mining companies apply just as much on the debt side. There are plenty of banks and other financial institutions who are now hesitant to provide debt funding to mining companies because of their 'green' lending policies. In many cases, this is simply pushing companies away from regulated debt into the arms of funders, often state-backed, who are quite happy to do business with miners but come with the same lack of transparency I have discussed earlier. This does nothing for sustainability and nothing for orderly financial markets.

Mining is hard work. The mining game is filled with uncertainty. It needs the support of funds and other institutional shareholders. These funds need to understand the essential nature of mining, and they need to understand that problematic assets are better off in the hands of responsible, transparent mining houses. They also need to support the mining industry as it provides the metals and minerals essential for a sustainable future.

~~~~~~~~~~

The final element to discuss in relation to mining finance is the vexed topic of taxation and royalties.

There have been a number of attempts over recent years to standardize many of the facets of business taxation around the world, including the ongoing work of the Organization for Economic Cooperation and Development (OECD) on BEPS (base erosion and profit shifting) and their efforts to encourage standardization of international tax rules more broadly through the cooperation of member countries. There has also been a significant strengthening of individual jurisdictions' transfer pricing rules, again with the aim of reducing opportunities to move profits from high-tax countries to countries with lower tax rates.

As you can't physically move mining operations, the greatest focus by tax authorities on the mining sector
~~~~~~~~~~

with regard to 'tax reform' has been on marketing and trading functions, with the best example being the commodity marketing teams often based in low-tax Singapore. This has led to a whole industry: aggressive claims by tax authorities of rampant profit shifting on one side; and the defence of their arrangements by the companies and their tax advisers on the other.

As with many plans that involve taxation, regulation and legislation, there are two sides to this story.

The tax authorities of countries with a higher fixed base of public expenditure, who also have a very developed mining industry, like Australia, often get very upset when they think miners are seeking to move profits to countries with lower tax rates. But the miners do this for at least two reasons, neither of which is born out of a specific desire to annoy legacy governments. Firstly, their primary duty is to their shareholders, their people and the communities within which they operate, all of whom benefit if the company pay the minimum global effective tax rate that they are obliged to. There's an additional, related benefit in the above-mentioned example of Singapore: the city-state has become a hub for commodity marketing functions and thus boasts the infrastructure, the pool of talent and expertise and a community of practice, not to mention a convenient time zone.

Secondly, in almost every case, companies are encouraged to move to low-tax jurisdictions by the inbound host governments, which see the upside in attracting businesses to boost their tax bases (albeit at a lower rate); increase the size of their economies; develop the service industries that support the inbound enterprises; and attract, to some extent at least, cohorts of well-paid, skilled workers who will then spend money and pay tax, contributing to the 'multiplier effect'. So when we see miners moving back-office and marketing functions to 'low-tax' jurisdictions, don't rush to judgement. There are sensible economic reasons for doing so. Plus, the noisy protest of the government that sees itself as losing out needs to be balanced with the usually much quieter approval of the government that has invited them in.

In addition to income-based taxes, the most common being a profit-based company tax, most countries apply a royalty on the value of production. The reasoning for this is straightforward: in many countries, the government, technically speaking, owns what is beneath the ground, so in exchange for allowing miners to dig it up, it charges a percentage on its value. In many respects, a royalty is the most logical of mining-related taxes, as long as it isn't abused. Which brings us to the next issue.

Mining companies are commercial enterprises which need to make a profit if they are to survive in the long term. So they will make commercial decisions about where to invest and operate. They acknowledge the need to pay taxes and royalties, but what they also need is stability and certainty with respect to the tax and royalty regime. Unexpected new taxes, or the sudden hiking up of the rates on existing taxes, won't endear a jurisdiction to those seeking stability and certainty as they make what are very long-term, very high-capital investment decisions. We have seen recent examples of this in recent years across Africa, South America and even Australia.

Mining is no different to any other industry with long time horizons. It requires stability and certainty in regard to taxes, regulation and government policy more broadly. It needs to know that a host government won't move the goalposts halfway through a ten- or twenty-year mine development project. As taxes and royalties can make a significant difference to the economic outcome of a mine, it's important for the integrity of the planning process that there can be a high level of confidence that taxes, royalties and compliance costs won't materially change over the course of its development and operation.

Recently, a new government in Peru, where the products of mining make up about 60 per cent of the country's export revenue, considered imposing 'super royalties' on the mining sector. Most of the miners with either existing mines or mines in development have what are known as tax stability agreements which give them certainty for a period of time. This isn't helpful, though, for companies considering new projects. The economics are difficult enough as it is without finding that you may have to unexpectedly hand over a much larger percentage of your revenue to the government. And it's important to make the distinction in this respect between revenue and profit. Corporate tax is (as a general rule) on profit. Royalties are on the value of what you mine, before taking into account what it costs you to do so. In most cases, you will still pay royalties even if, after taking off both the mining and processing costs and the royalties themselves, you end up losing money.

Other important elements of tax management to briefly mention include: understanding double tax treaties and how they will apply; having bilateral trade agreements in place that can be relied upon to stand up in court if necessary; ensuring that employment taxes are understood and appropriately factored in; the customs regime for getting material and machinery in and your metal out; and much more.

Finally, not a tax issue directly but another government-dependent factor that can have significant consequences for the profitability of a mining operation is the issue of capital controls. It's not very helpful if you find that your mine in a particular country makes a lot of money but then you are unable to repatriate those funds either to the country where your head office is located, or to the operations of your company in other countries where the cash is needed, or to shareholders by way of dividends. Sometimes capital controls are complete prohibitions from sending cash out of the country, and sometimes they are de facto controls by way of punitive taxation on any funds being sent overseas. Either way, this can have a significant impact upon your profitability and your cashflows.

Sensible tax regimes will encourage mining companies to invest with confidence for the long term, helping ensure we have those critical minerals and metals we need for a sustainable future.

~~~~~~~~~~

Mining is complex and uncertain. The very long-term nature of any mining project makes the economics difficult and the financing expensive. Miners have to navigate their way through the competing demands of shareholders, financiers, governments and other stakeholders to ensure their operations are able to be developed economically and then operated profitably. Don't underestimate how difficult this is. We need them to succeed, though, because we need the metal they produce.
~~~~~~~~~~

5

Why mining is essential for sustainability and the energy transition

We have spent the last few chapters considering what mining is, how it works and what its operational and economic challenges are. We now come to the central premise of this book: why mining is not only essential, but is becoming ever more essential as we seek to be more sustainable in the way we live, and in the way we produce and consume the energy we need to do so.

This premise may seem counterintuitive to many, particularly those for whom 'energy' as a general term still has associations primarily with burning fossil fuels. But nonetheless, the pursuit of cleaner energy demands more metal, and thus more mining. A simple example will help set the scene.

Take a wind farm. Harnessing the wind to generate electricity is one of the more straightforward renewable energy options. Although there are downsides – particularly the intermittent nature of wind, visual and noise pollution and the threat to birdlife – nonetheless wind is free, in the right places it blows almost constantly and the technology is well established.

A typical modern wind turbine will generate approximately 6,000 megawatt hours (MWh) of electricity in a year, assuming the turbine is turning for 80 per cent of the time. This provides sufficient electricity for 1,500 average households.[1] An efficient coal-fired power station uses 500 kg of coal to generate 1 MWh of electricity, so one wind turbine is saving 3,000 tonnes of coal from being burned.[2] Extrapolate that to an offshore wind farm of, say, 100 turbines, and you have saved 300,000 tonnes of coal and generated enough electricity to provide the power for 150,000 households.

But this wind-powered energy doesn't come free. There's a cost in terms of money and there's a cost in terms of the use of the earth's resources.

These modern, enormous turbines are built from a lot of steel and mounted in a lot of concrete. A single turbine, adding together the tower, the nacelle (the thing at the top of the tower holding the blade), the blade itself and the foundations, will probably be about 100 metres high, will have a blade diameter of at least 80 metres and weigh in excess of 300 tonnes. About 90 per cent of that weight is the aforementioned steel and concrete. That's a serious amount of steel and concrete, to which we need to add copper, various other metals and the fibreglass and resin for the blades. The 'glass' in fibreglass comes largely from mineral sand, but the resin that holds it all together is plastic, which comes from oil. As it turns out, even renewable energy needs oil. Using our example of a wind farm with 100 turbines, this renewable energy generator is made from over 30,000 tonnes of largely mined material. Wind farms also tend to be in remote locations such as windswept islands, or offshore, anchored to the seabed, so in addition to the metals and other materials for the wind farm itself, you need a vast amount of copper to send the electricity generated on its way and connect it to the power grid.

Wind farms are clearly becoming more and more efficient and are making a material and positive contribution to the energy transition. We cannot, however, underestimate their contribution to the ongoing and increasing demand for the mining of iron ore, metallurgical coal, copper, the constituent elements of concrete and the various rare earth elements required for their control systems and the magnets in their turbines.

This need for significant volumes of metals and minerals applies to all forms of renewable energy, and thus we return to the premise that the sustainability agenda makes mining even more essential.

Sustainability in general, and the energy transition in particular, bring significant change. With this change comes a focus on new materials, technological innovation and advances in the 'circular economy' and recycling, all of which impact on mining. These changes are, in principle, unquestionably positive: cleaner energy; technological improvement, which means we need less energy for the same outcomes; and better recycling, which means greater usage from the same amount of mined metal. This is all good. As a consequence of these developments, our stewardship of the earth's resources should improve, even with an increased dependence upon mining.

Let's explore the sustainability/mining nexus in more detail.

Renewable energy generation

I started this chapter with an example of renewable energy in the form of wind power. To this I would add solar power and hydropower. There are other, more niche forms of renewable energy, for example tidal power, but I won't cover those here.

There's no question that burning coal in the traditional way is polluting. You only need to look at photos of major cities like London taken before the mass cleaning of buildings that took place from the 1970s onwards. All the landmarks – Westminster Abbey, Buckingham Palace, St Paul's Cathedral, the Houses of Parliament – were black from centuries of airborne coal soot. The end of coal-fired heating and inner-city coal-fired electricity generation means that these buildings have remained (basically) clean. Extrapolate that to the pollution from a coal-fired baseload generator and you can understand why it's accepted that reducing or eliminating coal-fired generation is a good idea. We just need to remember that although we won't be digging up coal (or sucking up gas), we will be digging up lots of other minerals to enable the transition to renewable energy.

Wind turbines and solar panels are the two most significant contributors to the transition to renewable energy generation and, accordingly, to the changing demand for minerals. Advances in technology coupled with reductions in the development costs mean that many countries, including the UK and most European nations, now generate significant percentages of their electricity from renewable sources. Indeed, there have been windy days with low to average electricity usage when 100 per cent of the UK's electricity demand has been met from wind-powered generation. This is a good thing, although we still need alternative sources of baseload supply for when the wind isn't blowing, which on occasion it isn't.

We have already covered the material requirements of wind generation above, and it's a similar story with solar farms. Solar panels are installed in steel casings, with the panels themselves being largely glass, which comes from silica sand, lime and a few other ingredients. There's a further downside with solar energy generation: solar panels tend to work best in large arrays, and this means they're often installed on agricultural land. There's a limit to the amount of arable land which can be given over to energy generation without compromising food security. And again, there's the need for extensive infrastructure to connect the electricity from the solar farm to the grid. So it's important to recognize that wind and solar generation require a lot of infrastructure, and, of course, that comes almost entirely from metals.

There's a further element to renewable energy and that's the storage question. Because the sun is only up, on average, exactly half the day, and even then is often covered by clouds, reliance upon solar energy generation is vastly more attractive if the electricity can be stored efficiently. The same applies with wind power, given that sometimes the wind doesn't blow hard enough, and sometimes it blows too hard and turbines are stopped to prevent high-speed spinning from damaging the equipment. At the moment, the storage focus is on giant batteries. These require large amounts of lithium and other materials for the actual electricity storage, as well as steel, copper and other metals for the casings, housings, and the cabling to connect it all up.

The third renewable energy option that I want to briefly cover is hydroelectricity. Hydro has been around for almost as long as electricity itself and, put very simply, involves using the flow of water to drive a turbine and generate electricity. The water flow can either be from a fast-flowing river or, more usually, from a water flow created by damming a river and channelling water from the dam through a turbine.

The downsides of hydropower are significant, in that building a dam is not only very expensive and requires a huge volume of concrete and steel, but, more significantly, it requires you to turn a valley into a lake, with all the upheaval to ecosystems, livelihoods, properties and homes that will probably be required. Of course, the upside of building a hydroelectric dam is that you also gain a water storage facility which can then provide a reliable source of drinking water for downstream population centres.

Notwithstanding the downsides, hydroelectric generation infrastructure, once built, provides cheap, reliable and completely clean power and will do so for a very long time. Properly maintained, there need be almost no limit to the useful life of a hydropower plant.

Hydro often brings with it a very clever energy storage solution. Many hydropower plants include the capacity to pump water from a holding dam below the turbine to the main dam from whence it came. When demand for electricity is high, the plant runs as normal, with water released from the main dam, channelled in pipes through the turbines to generate electricity and then released into a holding dam. When demand for electricity is low, and there's unused capacity in the wider electricity grid which would otherwise go to waste, this surplus electricity is used to power pumps which send the water back up to the main dam, ready to be released and run through the turbines again at the next demand peak. This is very clever, as the marginal cost of the electricity to pump the water is nil, and you are thus effectively preserving your water supply for free.

In summary, wind, solar and hydropower are all excellent solutions, and indeed they are in widespread use already and will only become more prevalent. But remember, as demand for some minerals goes down (in this case, coal and gas), demand for many other metals will go up. My opening example of a modern wind farm is a case in point. The exponential growth in renewable energy will increase the demand for many key metals, and particularly the old favourites of steel, copper and the constituent elements of concrete. Even more effective recycling goes nowhere near satisfying this increased demand, and thus there will need to be more mining of primary metal. When taking into account the whole renewables life cycle, the environmental benefits of renewable energy are still very much there, but they come with trade-offs.

Nuclear power

Another technology in the energy transition mix is nuclear power. It's not strictly renewable, as a nuclear power plant consumes uranium. But nuclear generation produces zero emissions, is very safe, the plants last for a long time, they don't care if it blows a gale or there's no wind at all, and it makes no difference if the sun is shining or the rain is pouring. They make the ideal baseload electricity generation plant. So in determining what a balanced, sustainable energy future looks like, we should be looking harder at nuclear.

The difficulty with nuclear power currently is that plants take a very long time to build and they cost a fortune. The Hinkley Point C plant currently under construction in Somerset will generate sufficient electricity for about 7 per cent of the UK's needs, which is impressive. It will also cost over £30 billion (at current estimates) and is running about fifteen years behind schedule, which is perhaps not so impressive.

I mention earlier in this section that nuclear power is very safe. It hasn't always been so, but western nuclear plants that aren't built on seismic fault lines are totally safe. The International Atomic Energy Agency, the world's central intergovernmental forum for scientific and technical cooperation in the nuclear field, states that nuclear power plants are among 'the safest and most secure facilities in the world'.[3] It's also worth noting that the small amount of radioactive material, in the form of spent nuclear rods, is also able to be dealt with completely safely due to the protocols and infrastructure in place and its relatively insignificant volume.

The downside of this level of assurance is that it makes building a modern, western nuclear power plant incredibly expensive and time consuming. Nobody wants to compromise on safety, so hopefully a funding mechanism can be found that enables countries to build multiple plants, and the benefits of scale and replicability will kick in and the price per gigawatt hour will come down. Indeed, there's currently a lot of work going into the development of small modular nuclear reactors, which could unlock a future for economically viable, easily replicable, clean and sustainable nuclear power.

Where mining comes into this is that nuclear power requires uranium. If the world embraces nuclear power as a sustainable source of 'always on' baseload generation, we will also need to mine a lot more of it. Uranium mining production globally is currently about 50,000 tonnes per year. Over 40 per cent of this comes from mines in Kazakhstan, followed at a distant second by Canada on 15 per cent, then Namibia on 11 per cent and Australia on 9 per cent.[4] The World Nuclear Association estimates that global requirements are more than 60,000 tonnes, with the shortfall currently being made up from stockpiles of uranium held by miners, traders, governments and companies established specifically to warehouse it.[5] These stockpiles won't last for ever, though, and clearly new uranium mining will be required to deal with this supply deficit.

The final thing to note with respect to nuclear generation is the extent of mined materials required for the construction of nuclear power plants. A key element of any nuclear power plant is the 'containment vessel' within which the reactor sits. This containment takes the form of 'nuclear grade' reinforced concrete, which is very dense and able to contain radiation. This requires enormous amounts of cement and steel. As an example, according to EDF Energy, Hinkley Point C will use 3 million tonnes of concrete and 230,000 tonnes of steel (mostly for reinforcing) in its construction.[6]

So nuclear power, like all the sustainable energy opportunities we have discussed, may be emissions-free at the point of generation, but it will use a lot of energy to build, and need a lot of metals and minerals. Again, this isn't a reason not to use nuclear power, and as you have probably realized, I am a big fan of nuclear generation, but the metals and mining requirements need to be factored in.

Grid augmentation

I have previously mentioned the need to connect up new renewable energy sources to the grid and the impact of doing this on metals demand.

As we move towards the electrification of more and more processes, and add to that the capability of electricity grids to deal with renewable energy being fed into the grid from multiple sources such as domestic rooftop solar panels, the capacity of electricity grids needs to increase and they need to become more flexible. So we will need not just more generation, but also more transmission lines, distribution infrastructure, substations and 'smart' control equipment. This all needs more steel, copper and the other usual suspects of modern digital equipment. Added to the very metal-intensive nature of renewable generation, this all means a significant amount more metal, and yet more demands upon our miners to produce it.

It's also important to note the contribution to electricity usage made by our modern, connected world. Data on everything is available at the click of a key or a stab at a touchscreen, and all of that data being sent to and from 'the cloud', all those social media posts and all those control signals being sent via the internet require computing power to send them along their way. And this computing power comes in the form of massive server farms, or data centres, dotted around the world. These data centres already use about 2.5 per cent of the world's power, and this is only going to grow. There are countries which have made a business out of attracting and hosting data centres. Ireland, as an example, as of the end of 2023 hosted eighty-two data centres, according to the *Financial Times*. Those eighty-two data centres used 21 per cent of the country's total electricity generation capacity.[7] This is an astonishing amount of electricity. Generating it, transmitting it and connecting it to the data centres requires even more copper and steel for the generating plants, the transmission lines, the substations and everything else. This is something we should keep in mind when we think about how clever it is that we can control all of our supposedly sustainable gadgets remotely. Another point to note is that, as anyone who has sat with a laptop actually on their lap for too long will know, computing generates a lot of heat. So a huge server farm requires a lot of cooling, and that uses a vast amount of water, twenty-four hours a day.

The recent development of AI – artificial intelligence – and its widespread use in both domestic search engines and commercial applications is only exacerbating the demand for energy and water, and thus ultimately metal. A recent article in the *Financial Times* quoted a report from the International Energy Agency (IEA) which noted that online searches using AI consume ten times the electricity of traditional searches. According to the IEA, as a consequence, the global electricity demand from data centres in 2026 will be double what it was in 2022, an increase 'equivalent to the total power demand of Germany'.[8] Keeping those data centres supplied with electricity is going to need a phenomenal amount of metal.

The power, water and infrastructure requirements of the internet and the cloud aren't reasons not to embrace them, but just more of those trade-offs that we have to make. You want connectivity, you want the ability to access all your information, everywhere, all the time? You will need more power generation, which requires more metal, which requires more mining.

In summary, the transition to more sustainable, renewable energy is an excellent endeavour. Combine this with greater efficiency so we consume less energy in the first place, and then add in more effective recycling (a key element of the 'circular economy'), and we are able to say with confidence that we are heading for a more sustainable future. And to get to that sustainable future, mining will be essential.

Electric vehicles

Another significant and well-publicized source of new demand for metals is electric vehicles (EVs). There has been an almost religious fervour around EVs in certain circles, and I applaud both the innovative thinking that's at work here, and what is clearly a genuine and authentic desire to find an alternative to petroleum-driven motor vehicles.

But in the rush to embrace battery EVs, I urge caution. The goal of end-use 'purity' may make the drivers of EVs feel good about themselves as they drive almost silently along, burnishing their green credentials, but the benefits may not be as clear-cut as they think. A good example is aluminium. Traditionally, car bodies were made of steel. In an effort to make them lighter and more fuel efficient, and, in EVs, to compensate for the weight of battery packs, an increasing proportion of vehicle bodies are now made from aluminium. Aluminium is made by smelting alumina, which in turn, as we have noted above, is refined bauxite, and the whole process of making the metal uses a very large amount of electricity. So much, in fact, that one of the nicknames for aluminium within the industry is 'frozen electricity'.

Another metal which is central to EVs, as it is to anything with the word 'electricity' in it, is copper. Depending upon the model, an EV has three to four times the amount of copper in it compared to an internal combustion engine car. The supply of copper is perhaps the most fundamental issue for everything to do with renewable energy and the electrification of both mobility and industrial processes. This is because not only do electric motors and batteries use a lot of copper, but we also need copper for the wiring and cabling to get the electricity to the point of use or, in the case of vehicles, the point of charging. Then we need copper for the generation. As you can't have electric-powered anything without first having electrical generation, there's a 'multiplier effect' on the need for copper.

There are very valid concerns about the ability of the mining industry to keep up with the anticipated demand for copper in the decades to come. BHP, the world's biggest miner, noted recently that it expected the world to need an additional 10 million tonnes of copper over the next ten years, rising to forecast demand of 50 million tonnes per annum by 2050.[9] That's double current new copper production, which was 25 million tonnes in 2023, to which was added 6 million tonnes of recycled scrap copper. About a quarter of this total is copper for use in energy transition applications, including EVs and the additional infrastructure and generation capacity required to charge them up. BHP's analysis is broadly backed by most analysts and industry commentators.

The problem is that many of the world's big copper mines are getting old. Grades are reducing, which means less copper is being produced for the same amount of mining. Additionally, as analysts have also noted, almost every mooted new copper project either is running very late, or has been put on hold, usually because the cost of financing and building them has become prohibitive. The consequence is a potential supply deficit estimated at up to 20 per cent of demand by 2050. The energy transition cannot be achieved without copper, which means, if the world is to have all the copper it needs, it will have to pay a significantly higher price for it to make it economically viable to build all the expensive new mines we will require. Another trade-off on the road to sustainable mobility.

Turning to the batteries powering EVs, there are a few things to consider. The chemistry of an EV battery can vary, but in most of them, the key metals are lithium, manganese, cobalt, nickel, graphite, aluminium and copper. The sourcing of some of these metals has been contentious, particularly cobalt, where the supply chain can be opaque. The same applies to rare earths, which are another important component of EVs, smartphones, wind turbines, military applications and much else. The major supplier of rare earths continues to be China, not exactly a standard bearer for transparency and disclosure. Added to this is the concern that extracting rare earth elements from their host ores is very water-intensive and produces a significant amount of pollution, including heavy metals, in the production process.

The batteries of EVs also have a limited life. It's generally considered that after ten years they will be degraded to the extent that the host vehicle will have a seriously compromised range. Given that range anxiety is one of the major reasons private buyers have not embraced EVs to the extent that was expected by many forecasters, this is a limiting factor for the life of an electric car. As it isn't really practical to replace the battery pack in current-generation electric cars, the car itself will need to be retired. Fortunately, the capability and market for recycling battery packs are growing quickly. It's estimated that by 2030 there will be well over 1 million EV battery packs available for recycling each year, a number that will only increase. Notwithstanding this recycling, if the EV market grows as predicted by many forecasters, the demand for 'new' lithium, nickel, cobalt, graphite and other metals will expand strongly, and that means more mining.

The final issue to consider with battery EVs is the contentious issue of charging. This is the one that could scupper the whole enterprise in its current form. Do you remember the fax machine? They were brilliant and changed the way we communicated and did business for ten to fifteen years, and then they were gone. Made completely redundant by technological advance. Are we absolutely certain battery EVs won't turn out to be, ultimately, a transitional technology? Will they be the fax machine of the automotive world? The intractable problem with battery EV charging gets remarkably little airplay. To use the UK situation, but one which will apply equally across any city in the world with high-density housing: approximately 30 per cent of UK homes don't have off-street parking. There are approximately 30 million vehicles in the UK, so that means about 9 million of them live on the street. Creating practical charging infrastructure for these 9 million vehicles is, I would suggest, unlikely to be achievable. An innovative solution will be required.

Suffice to say there will need to be an awful lot of clever innovation and technological development if the use of this type of vehicle as a means of regular mass transport is to become in any way practical. If battery EVs really do become the default domestic motor vehicle, then we need to accept that there's currently no practical solution to the conundrum of long charging times, the lack of infrastructure, owners without access to off-street parking and the electricity grid augmentation that will be required. So, hopefully, the required innovation, technological development and capacity investment will happen. And whatever form this takes, it will need more metal.

The wider transport question

We have talked about the 'energy transition' largely in relation to EVs, but that's only the start. We need to consider all the other transport-related questions, and particularly trains, aeroplanes and ships. The electrification of railways has been a thing for well over a century, and electrically powered locomotives and multiple units are normal and will no doubt only improve in efficiency over time. But, once again, as railway operators around the world move more and more from diesel to electric traction, there will be a growing need for all the metals required to manufacture the locomotives, multiple units, catenaries, substations, transmission lines and ultimately increased generation capacity.

Sustainable aviation is in its early stages, although it's important to note that the modern aircraft engine is much more fuel-efficient than the first generation of jet engines. We may not have a viable alternative to aviation fuel (often called 'avgas') at this stage, but the aviation industry is working on it. The problem that needs to be overcome is lack of energy efficiency. For an aeroplane to be able to fly any meaningful distance, the weight-to-energy ratio needs to be low. Battery-powered propulsion is simply too heavy to enable sufficient batteries to be carried to provide sufficient range. Perhaps the solution is rather in synthetic fuels?

The big 'non-EV' transport fleet is shipping. In a globally connected world, there's a vast fleet of ships, both bulk carriers and container vessels, sailing the world's oceans delivering raw materials and finished goods from producers and suppliers to end-users and customers. At last count, there were over 13,000 bulk carriers and 6,000 container ships,[10] and that's before we come to the vast number of other, smaller cargo vessels, which run into the many tens of thousands. And they are all made of steel. Their engines run, almost entirely, on heavy fuel oil, which is a particularly unpleasant, thick form of oil. One of the consequences is that ships running on heavy fuel oil produce relatively high levels of pollution from their exhausts. There are now a small but growing number of ships which run on gas, a much less polluting hydrocarbon, so the shipping industry is doing its part, but there's a long way to go.

The thing that makes the place of shipping in all this such a paradox is that achieving our sustainability goals, as we have noted (repeatedly), requires all this extra metal, which has to be shipped around the world on, you guessed it, more and more bulk carriers. All that essential metal, mined in one place, processed in another place, turned into manufactured goods in a third place and sold and consumed in a fourth place, requires a lot of shipping. That means more steel and everything else, and that means more mining.

Hydrogen

No discussion of sustainable energy is complete without raising the issue of hydrogen. Hydrogen has great potential for powering vehicles, particularly larger vehicles where there's capacity to design in a large hydrogen tank to supply the fuel cell. By way of example, haul trucks running on hydrogen are already an early-stage reality, working in South African platinum mines. These haul trucks are demonstrating not only the viability of hydrogen as a fuel source, but also an alternative use for platinum, which is used for the electrodes in the fuel cells.

There are two concerns commonly expressed with regard to hydrogen as an alternative to battery-electric power for vehicles. The first is safety. I think too many people have in the back of their minds the *Hindenburg* disaster of 1937. They need to stop equating modern hydrogen storage technology with an enormous airship full of hydrogen constructed with 1930s technology almost one hundred years ago. The management of hydrogen in the applications being developed today doesn't present any particular safety concerns. And, on that subject, have you seen what happens to a battery EV when it catches fire?

The second concern, which is more valid, is the enormous amount of electricity required to make hydrogen, or, to be more specific, to separate hydrogen from wherever it's in its natural state, most commonly water. In other words, to get rid of the O in H_2O through electrolysis. Understandably, those who are promoting the use of hydrogen fuel cells are very engaged in developing ways to make hydrogen in as green a manner as possible. The platinum mines I mention above are doing this by installing large-scale solar generation.

Planning and developing the necessary additional electricity generation to go along with hydrogen production will be essential if this technology is to work, as we are talking seriously large amounts of electricity. And once again, more electricity generation means more metal. A technology receiving a fair amount of press, including in discussion papers from the European Union, is the potential use of hydrogen to make steel. Instead of using coke (which comes from metallurgical coal), use hydrogen instead. This sounds great as, unlike coke, burning hydrogen results in zero emissions. But first, as noted above, you have to make an enormous amount of hydrogen. And this enormous amount of hydrogen will require an equally enormous amount of electricity. To put that in context, one quote suggested that a plant to manufacture steel using this method that's being proposed for Sweden would require electricity equivalent to one-third of that country's total electricity generation capacity.

None of this is a reason not to pursue hydrogen as a fuel. But it's a reason to ensure we always consider the whole life cycle of energy use before making decisions. And that includes the increased requirement for a significant range of metals to enable hydrogen technology to flourish.

Heating

One of the biggest sustainability-related issues of the moment, particularly in colder climates, is heating. Most heating in the UK is carried out with gas-fired boilers sending hot water around houses to heat radiators. There's a desire on the part of policy makers to reduce our dependence upon gas and look at electrically powered heating solutions. The one being talked about at the moment is the 'heat pump'. Currently, heat pumps are noisy, inefficient, expensive to install and heat water for circulating through a building's radiators at a lower temperature than traditional gas boilers. They are at an early iteration of their development, though, so perhaps after a few generations of heat pumps and technological advances in this area, they will be a viable mass-substitution solution. Let's wait and see. If they are, there's even more electricity required, even more grid augmentation, and thus even more metal.

All the above may start to sound a bit repetitive, and that's because it is. A sustainable future requires changes to the way we generate and consume power, our mobility, our communications, our heating and every aspect of our lives. We went from inventing the aeroplane to putting men on the moon in sixty-six years, so I have great faith in the capacity of the human race to find technological solutions to enable our quest for sustainable energy, sustainable mobility and, indeed, sustainable everything. We will solve the various energy conundrums we face, sooner or later, even though we may not yet know exactly how. One thing I do know is it will involve a lot of metal.

Agriculture

A final element of the sustainability and mining jigsaw is in relation to what we eat. We couldn't have modern agriculture, and thus the capacity to feed the world's population, without the essential and significant contribution of metals and minerals. In the ultimate expression of sustainability, mining enables the sustaining of life through the production of sufficient food.

It does this in two ways. Firstly, the mining of potash and other minerals for the production of fertilizer enables much more productive farming. The use of minerals-based fertilizers has more than doubled the productivity of agricultural land. Farms, orchards and market gardens, among other sources, are able to provide the world with the food it needs. This increased productivity also means the footprint of agriculture is reduced, enabling more land to be left in its natural state with all the sustainability and biodiversity benefits that brings.

Secondly, modern agriculture uses machinery for cultivation and harvesting; silos, buildings and other structures for storage; machinery and factories for processing and packaging; and vehicles and infrastructure for transport and distribution. All of these things are primarily made from metals and minerals.

It's tempting to say that we could do without a lot of that if we just ate more 'natural' foods and steered clear of ultra-processed food. There's certainly a very good case for reducing our intake of ultra-processed foods. However, even simple food production often requires a lot of equipment. Think of milk and dairy production, abattoirs, bakeries, poultry and eggs, even something as simple as the sorting and cleaning of vegetables for market. These all use facilities, buildings and infrastructure made from metals, mostly steel. Once again, substituting out metal, for example for holding tanks, will only mean the replacement of metal tanks with plastic ones. Mining or oil, take your pick.

So sustainable agriculture needs mining, and a lot of it.

Recycling

People sometimes ask me, 'Instead of more mining, couldn't the solution be an increase in recycling?'

Recycling has been a thing for ever, and there's nothing new about reusing materials. To achieve our sustainability goals, we need to get better at recycling, and we need to find ways to recycle metals which currently are considered difficult or impossible to reuse. It's important to note, however, that recycling is a contribution to meeting sustainable metals demand, but it comes nowhere near replacing the mining of primary metals. This is for a combination of two reasons: the demand for metals is increasing much more quickly than the availability of secondary metals from recycling; and exacerbating this, a significant proportion of total metals consumption is in long-life assets like buildings and infrastructure. This means metal is 'locked up' for a long time, while demand for new metal continues.

Interestingly, as much as we want to recycle as high a proportion of scrap material as possible, we also want to ensure the things that we make from metal last as long as possible, because replacing them uses energy and resources. So if we need more primary metal because all the things we use metals for are lasting longer before needing replacing, then that's a good thing.

An excellent example of the benefits of recycling in relation to sustainability is steel. Steel is easily recycled, and indeed can go on being recycled for ever. The more scrap steel there is available, the more new steel can be produced in electric arc furnaces, rather than blast furnaces, thus reducing the need for coke production from metallurgical coal, a process which, with the best will in the world, we need to acknowledge is quite polluting.

According to the European Union Circular Economy team, over 35 per cent of total crude steel production is from 'secondary material', that is, from recycled steel. This is important because recycling uses 72 per cent less energy than primary steel production. In terms of metals and minerals, 1 tonne of steel recycled saves 1.4 tonnes of iron ore and 0.8 tonnes of metallurgical coal.

Finished copper is also eminently recyclable, as is aluminium. However, not all metals in their end-use state are. One of the biggest problem areas at the moment is how to recycle rare earth elements and the other metals in mobile phones, batteries, IT systems and many of the other high-tech applications in use. As battery EVs, with batteries that degrade over time, and battery banks to store intermittent renewable energy become more and more common, the ability to efficiently and effectively recycle the 'black mass' becomes ever more important. As these metals are also among the more difficult or more expensive to mine and process, the ability to efficiently recycle them is even more pressing.

The technology around recycling will doubtless develop, and will probably develop quickly, but it will only ever provide a portion of the metals we need. We will still need a lot of mining.

6
The mining industry's role

If mining is essential to our transition to a more sustainable future, then clearly mining companies will play a major role as our societies and governments grapple with environmental challenges over the next few decades (and beyond). Mining companies need to improve their own environmental performance, but just as crucially, governments and investors are going to need to be realistic and work constructively with them to ensure that we get the minerals and metals we need, because, as we have seen, mining is a very complex and challenging business that can easily be adversely affected by a vast range of factors. How can governments, investors, non-governmental organizations (NGOs) and various other stakeholders, and the mining industry itself, create a way forward that's both responsible and realistic? That's the subject of these next two chapters.

The external view

Firstly, let's consider mining and the broader economy; perhaps we could call this the *external view*.

We have discussed at some length the increase in demand for metals to meet the needs of the sustainable energy transition. Every aspect of the transition to renewable energy requires more metal. Electrification of everything requires vastly more copper. Renewable energy installations require more steel, more copper, more aluminium, more rare earths, more lithium, more nickel, more glass and thus more mineral sands. It's the mining sector's role to produce these commodities in a sustainable way, but produce them they must, if we are to achieve even some movement towards the goals being set, particularly the 'net zero' goals currently being discussed. It's the role of governments, activists and investors to take a pragmatic, more thoughtful approach to mining. They should be promoting responsible, sustainable mining, instead of pursuing knee-jerk anti-mining policies that will inevitably turn out to be counterproductive for both the mining industry and the cause of sustainability. They should be as helpful as possible, doing all they can to smooth the way for miners to develop new operations and bring on stream the necessary increases in production. I suggest a change of mindset may be required.

This is illustrated particularly clearly with regard to the concept of 'scope 3 emissions', the notion that a miner should take into consideration the emissions which will be caused further down the supply chain by their customers, or indeed by their customers' customers. It's the reason many miners have decided to divest their thermal coal assets. Unfortunately, simply 'getting out of coal' will do precisely nothing for decarbonization. All it will do is move coal-producing assets out of the public gaze and enable virtue-signalling investment funds to declare their green credentials.

Indeed, demanding that listed mining companies should divest, as quickly as possible, from coal mining assets, and particularly thermal coal assets, is one of the more poorly thought-through campaigns of eco-minded activist investors. Although the view that burning coal is bad for the environment clearly has merit, the campaigners need to be careful what they wish for. Listed miners are accountable to their institutional and retail shareholders, to regulators and to the overarching court of public opinion played out in the business and general press. They are very aware of the issues around mining and marketing thermal coal and are alert to the need to manage down their coal portfolios in an orderly manner. In the hands of responsible miners, this will happen.

But, as noted in chapter 4, a fire sale of coal assets may, and often does, result in ownership being transferred from responsible, heavily scrutinized public companies to opaque, unaccountable private or state ownership. The mines won't be wound down, but will continue, because the demand for thermal coal continues to be huge. Recent geopolitical uncertainty, the impact upon the reliability of gas supplies and the ongoing increase in demand for electricity is only making the demand for thermal coal even greater. In fact, a record 8.74 billion tonnes of coal was produced globally in 2023.[11]As I write this, there are over 600 new coal-fired power plants either planned or under construction around the world. So coal-fired generation won't stop. With the transfer of ownership the activists are demanding, what *will* stop will be meaningful oversight of the coal mining required to supply these plants. The new owners won't care what the activist investors, the environmental campaigners or the NGOs think. They will have acquired operating assets, probably at bargain prices, for which there's a massive market in China, India, the rest of South East Asia, Africa and South America. To these markets I add a lot of North America, Europe and Australia, despite the public perception that coal-fired electricity generation is a thing of the past in 'the West'. Germany, for example, often held up as a bastion of renewable energy generation, still produces just under a quarter of its electricity needs from coal, a position made worse by its rushed and poorly thought-through decision to get out of nuclear generation. So in other words, there's a continuing thermal coal market just about everywhere. The difference is that there's increasingly less accountability. So perhaps a more orderly approach to dealing with thermal coal is in order, and perhaps the mining sector needs to be better at explaining these sorts of issues too.

Of much more use have been those instances where thermal coal mines have been demerged into entities, still listed on well-regulated exchanges, that can continue to produce to meet the very real current demand while at the same time looking at ways to improve sustainability and find alternative, cleaner options in the medium to long term.

For those who are determined that miners must take responsibility for emissions further down the supply chain, again I would quote the adage 'be careful what you wish for'. If the miners are to be held responsible for other people's environmental damage, should they also be responsible for customers' disregard for human rights, or their disregard for democracy and the rule of law? Given that much of the world's iron ore and copper is sold to state-owned, or at least state-controlled, entities of authoritarian regimes, it seems to me that this should be causing just as much concern as what happens when they burn the coal the miners sell them.

A further consideration when assessing the miner's role in our sustainable future is the issue of sourcing critical metals and bolstering our resource security. Ultimately, we need governments, miners and broader industry to work together constructively to enable the development of new mining and processing capacity to wean us off our dependence upon one country for critical minerals, metals and essential components.

This is a very important consideration as the processing of most rare earth elements (REEs) used in everything from advanced componentry, electric motors, computer chips, touchscreens and many other applications is currently done in China.

A few years ago, western governments woke up and finally realized they had a security-of-supply problem.

With increased geopolitical uncertainty and growing concern around the dominance of individual countries over metals supply chains, the issue of critical minerals has come to the fore and the result is the current rush by many governments to implement 'critical minerals strategies' and associated policies. This is a good thing, as long as all the critical minerals talk turns into actual mining and, more importantly, actual processing. The development of alternative sources for these vital raw materials and components will be necessary if we are to confidently rely on sustainable energy and a lot of other new technologies in the long term.

For a long time, the world was quite happy to allow China to dominate the processing of REEs. This was partly out of apathy – the Chinese probably realized earlier than anyone else the value these rare earths would have in the years to come. But it was also out of self-interest. Mining and processing REEs is a fairly disgusting process. On average, producing 1 tonne of them leaves behind 75 tonnes of toxic waste water, as well as dust, waste gas and radioactive residue. So the major economies of the West, with their very active domestic environmental lobbies, were quite happy to avoid all the mess by leaving the mining and processing to China.

The problem is that these critical metals and minerals are vital to the energy transition. For example, neodymium is essential in the manufacture of permanent magnets, without which you can't have a wind turbine or, for that matter, an electric motor, which is exactly the same thing, just operating in reverse. Yet practically every permanent magnet comes from China, because it processes almost 100 per cent of neodymium. It's quite sobering to realize that nearly every vehicle, machine, appliance and wind turbine in the whole world relies on one country for the supply of the electric motors that enable them to operate.

Another example is the manufacture of touchscreens. Touchscreens require a number of REEs, particularly indium. The touchscreen is now all-pervasive, used on

every smartphone, in most motor vehicles and on almost every control panel on any piece of machinery or appliance anywhere. So once again, if a country wishes to manufacture touchscreens, it needs to ensure there's a reliable and transparent supply chain for the relevant REEs.

Other critical minerals examples include high-purity manganese, necessary for batteries, and even nickel, which is also a fairly unpleasant metal to process. Again there's an over-reliance upon one country for the supply of these metals: South Africa in the case of manganese and Indonesia in the case of nickel. It's worth noting, though, that neither of these countries has the almost total dominance of China with neodymium.

The common element in all these examples is that through a combination of poor strategic planning, short-term environmental decision making and, until recently, a lack of joined-up thinking from governments, particularly in the West, we now find ourselves almost totally reliant upon one country for the supply of a whole range of minerals and metals which are vital to the energy transition that we all say we want.

In the UK, for example, responsibility for policy in relation to the mining industry is spread across three different government departments. Add to this an almost irresponsible level of ignorance about the industry, and what you get is a policy vacuum and no strategic direction. Happily, the UK government has now come up with a critical minerals strategy. It remains to be seen whether this translates into critical minerals action.

It isn't just rare earths either. The broader issue of resource security is becoming much more urgent.

In many western countries, there's an over-reliance upon imports of essential minerals, not necessarily because there are no commercially feasible deposits in-country, but rather because there's no political will to make the business of mining as easy as possible through streamlined regulation, sensible permitting, a responsible but sensible approach to environmental matters, development-friendly

taxation and a coherent policy framework. Consequently, it's all too hard, too expensive, too frustrating, too difficult to get past the various interest groups, and too time consuming, so the mines don't get developed and the imports continue.

Perhaps, most of all, there needs to be a concerted effort to overcome the stifling 'nimbyism' ('not in my back yard') which kills off so many mine developments in developed countries. This is made worse by often illogical environmental campaigning, by people who say they want reliable, ethical and sustainable energy but refuse to countenance the very developments which would enable reliable, ethical and sustainable local sourcing of the necessary raw materials.

Resource security and a critical minerals strategy are vital to ensuring we maintain a reliable energy supply, reduce our dependence upon imports for both raw materials and manufactured goods, ensure we aren't beholden to potentially hostile states and can peacefully maintain our standard of living.

To achieve all these things, there will necessarily be trade-offs. The environmental absolutists will probably hate that, but we mustn't be deterred. Implementing that critical minerals strategy will require a grown-up approach to mining.

On a more positive note, another external impact the mining industry can have is to be active in research, discussion, innovation and development of new uses for the metals and minerals they produce. This will usually be as partners in the broader development of more sustainable energy, of more environmentally friendly construction and manufacturing through innovative end-products and more sustainable agriculture. There are already many examples of this, including through the work of bodies like the International Council on Mining and Metals, or metal-specific organizations such as the International Copper Study Group, based in Lisbon. Then there are the many academic bodies, including the Camborne School of Mines,

the Colorado School of Mines, the Sustainable Minerals Institute of the University of Queensland and many similar schools around the world; the research institutes partnering with mining companies; and the various mining-specific conferences such as Mining Indaba in South Africa and the PDAC (Prospectors & Developers Association of Canada) conference in Toronto. Put together, these advocacy, education, research, innovation and development bodies make a significant contribution to the better and more sustainable use of metals from mining.

Some examples where the mining industry is working with broader industry, academia and governments on more sustainable outcomes from metals include the development of hydrogen-fuelled vehicles, trialling new, lighter and more efficient types of building materials, seeking alternatives to coke in steel making and working on less carbon-intensive cement, currently one of the most polluting of all building materials. Spending management time and shareholders' money on ways to use metals for a more sustainable future is a genuinely valuable contribution mining companies can make, and much more productive than arbitrarily deciding who is green enough to be allowed to buy your product.

The internal view

Secondly, the mining industry is also working to address sustainability 'within the mine gate', which we can call the *internal view*.

The industry's achievements in this area are many, although almost totally unheralded outside the sector. One of the biggest enablers of more sustainable mining over recent decades has been astonishing technological advance. In addition to the sustainability dividend, this has led to remarkable and sustained improvements in safety, productivity, efficiency, environmental impact, community engagement, working conditions and much else. This is

an area in which the industry has made great strides and can take justifiable pride, even if almost no one outside the industry would be aware of it.

Just some of the technology-led sustainability-enhancing advances include the electrification of plant and equipment previously operated by diesel or gas; innovative and more accurate ore identification technologies which significantly reduce the amount of blasting and shovelling required; the adoption of in-pit processing to reduce the amount of material movement required; and remarkable reductions in water usage.

Perhaps one of the more game-changing technological innovations has been mass digitalization, most notably through the use of sensors on just about everything, to enable real-time adjustment to operations, effective predictive maintenance and the successful implementation of remote operations, significantly improving productivity, efficiency and thus sustainability. Another very important consequence of mass digitalization is removing people from the mine face, with a direct improvement in safety. So effective is the digitalization of mining that some of the world's largest mines, the Pilbara iron ore mines operated by Rio Tinto, BHP and others, are operated in all material respects from 'remote operations centres' in Perth, about 900 miles from the mines they are controlling.

These are all activities currently being undertaken by mining companies to improve sustainability outcomes within their operations, that is, 'inside the mine gate'. Many of these are in operation already, every day, in mines all over the world, and many more are under development.

And they aren't happening in a bubble. Mining is wisely borrowing best-practice ideas from adjacent industries. It's using the skills and innovations of the IT industry to apply to the sensing, data collection and analysis opportunities of digitalized processes; it's reliant upon the developments in the renewable energy sector to power operations; it works with the water industry around desalination and

recycling; it works with transport and logistics providers; and it works with the motor vehicle industry on its mobile fleet.

To explore in a little more detail one area where the mining sector has been at the forefront of technological innovation, let's look at digitalization – the use of sensors to collect previously unimaginable volumes of data, in turn enabling the management and control of mining, transporting, maintenance and processing tasks with a level of precision not previously possible. Digitalization has often gone hand in hand with other technologies: for example, driverless haul trucks and even, on some of the Australian iron ore railways, driverless trains. The notion of a mile-long train, loaded with thousands and thousands of tonnes of iron ore, rolling through the Australian outback with not a soul on board may be a bit unnerving to contemplate, but with the technology the mining companies have developed, it's a safe and very effective reality. Digitalization also assists with electrification, where previously diesel-powered processes are converted to electric power, enhancing sustainability. I guarantee you most people wouldn't associate mining with the cutting edge of the digital economy, but they should.

It's important for miners to flag up their embrace of innovation, because the general view of mining, from my experience, is that it's old fashioned, dirty, inefficient and incompatible with the digital economy. Thus, in many people's minds, it's equally incompatible with a sustainable economy. As it's becoming both digital and sustainable, the industry needs to get the message out there.

All the above contribute to a reduced environmental footprint inside the mine gates, and, as the environment doesn't have walls, by extension they are contributing to sustainability outcomes beyond those mine gates too.

One of the concerns some people have about mining is what they would regard as the visual pollution of a mine, so having spent a few paragraphs talking about all the good things miners are doing 'within the mine gate', we

should address the issue of those enormous holes in the ground inside those gates.

Mining often looks bad, quite literally. When you get up close and personal to a mine site, it really isn't pretty, consisting of very large holes and usually a lot of dirt. In fact, a universe of dirt. Ore stockpiles; waste dumps; plant, facilities and buildings which have been built 100 per cent for function and 0 per cent for form. There are conveyers rising and falling seemingly for no particular reason and often a lot of what look like randomly discarded bits of equipment and machinery. In the middle of a desert, this may not be so bad. In the middle of a tropical forest, it can be quite jarring.

However, less than 1 per cent of the world's land mass is used for mining, and a vastly smaller percentage of that is actual 'holes in the ground'. This is pretty minimal for something so essential. In any case, most mines are in remote locations such that unless you make a specific effort to go there, you will never see them, and as long as the pollution is only visual, they shouldn't upset anyone.

It's my experience that most people who are rude about the visual pollution caused by mines have never been near one. They have probably seen photographs or videos of mines which, frankly, have often been framed to maximize the ugliness and the sense that they are a blight upon the landscape. It often reminds me of the photos and videos used in the western press whenever there's an article about the pollution caused by traditional power generation of any kind: coal, gas or nuclear. They invariably show a line of concrete structures that look like huge salt and pepper shakers, with clouds of white stuff coming out of the top. What you are meant to think is: 'Look at all that horrible polluting smoke coming out of those enormous chimneys.' What you are actually looking at is zero emission, totally harmless water vapour rising from the top of water-cooling towers. But the average reader or viewer doesn't know that, and the damage to perception is done. Both the mining industry and the

energy industry need to get a bit more in control of their narratives.

People like me actually find a large open-pit mine a thing of beauty. They are huge, they have been constructed with a lot of care, thought and planning and, contrary to popular belief, they are often the perfect blend of form and function. From above, the 'benches' and the haul roads intersecting and diverging make them look like a gigantic M.C. Escher drawing. And when the pit is juxtaposed against the Andes mountain range, or against the endless nothingness of an Australian or Chilean desert, or with the backdrop of an expanse of snow-covered Russian wilderness, it's very impressive. But I accept that people like me are in the minority outside the industry.

The industry response

This is no time for complacency, so what else could the mining industry be doing both to promote sustainability itself and to ensure it can get on in peace with the job of providing all the metals for the energy transition? I am sure there are many things, but three come to mind.

Firstly, the mining industry is by no means perfect. For all the great work being done, and noted briefly above, there are unfortunately still miners who are environmentally destructive, wasteful and careless. There's also a middle group who are seeking to improve, or at least seeking to look like they are seeking to improve, but who show little sign of dealing with the legacy issues of decades of environmental degradation wrought upon the landscape, and sometimes upon the communities, around their mines. So there needs to be more effort put into bringing the companies who own such operations to accept the need for, and genuinely implement, reforms and improvements. This won't be easy, particularly when mines are operating in jurisdictions which themselves don't really care about the mess their resident

miners are making. I discuss this some more in the next chapter.

But it's important that the responsible end of the mining community do all they can. Demonstrating a commitment to cleaning up the laggards of the industry, in addition to their own efforts, will go a long way to convincing the broader community, including investors, regulators, politicians, lenders and activists, that the mining sector is serious about its commitment to sustainability.

Secondly, the mining industry needs to get a lot better at explaining not only its own essential nature, but also the work it's doing. We often hear the expression 'actions speak louder than words'. In the case of the mining sector, the actions are happening but few people see them. The world is largely ignorant about the contribution of the mining industry to both their daily lives and the sustainability agenda. I normally recommend against undue self-promotion, but in this instance I believe a bit more of this is necessary.

Thirdly, there needs to be consideration about how to persuade recalcitrant governments to implement policies in their domestic mining industries to improve both environmental management and sustainability. Without wanting to sound overly cynical, I do sometimes despair at the western liberal tendency to participate in talkfests like the various COP (Conference of the Parties) meetings and then triumphantly announce some fabulous planet-saving agreement or other, only to express surprise over the following years when most countries haven't implemented the policies and actions they promised. Why on earth are they surprised? These multilateral meetings provide wonderful forums for virtue signalling, grandstanding and Olympian levels of disingenuity. The well-known expression 'insanity is doing the same thing over and over and expecting different results' is often misapplied, but in this case I think it may be appropriate.

If you actually want to get change happening, history tells us that direct talks (usually undertaken away from the spotlight), with encouragement from major employers, major taxpayers, major customers and major funders, will go much further and produce more tangible, lasting results. Governments tend to respond to threats to their tax base and threats to the GDP growth they have promised their constituents much more than the short-term, quickly forgotten fanfare of the communiqué at the conclusion of the latest climate conference. The mining companies themselves have a crucial part to play in this, being both major companies in their own right and, in many of the countries we are talking about here, significant employers, taxpayers, suppliers of materials and generators of foreign exchange. They thus have serious influence which they should use for good.

The world is going through a fundamental energy transition, reducing reliance upon polluting fuels and moving to greater and greater use of clean energy. Miners are absolutely essential to this transition, both through the supply of materials required to achieve it and through their own sustainability accomplishments.

Finally, on mining and sustainability

A word of caution. Some of the more zealous climate campaigners insist upon what one might call 'absolute' outcomes. It could be argued that the more passionate 'net zero' zealots fall into this category. They want zero this and the total elimination of that. I suggest they are being unrealistic, unhelpful and indeed counterproductive to the cause of sustainability.

If we wish to benefit from the good things we take for granted, the improvements gained over hundreds of years of technological, scientific and social advance – comfortable, pleasant and practical housing; widespread mobility; convenient and instantaneous communication; universal connectivity; accessible and affordable consumer goods; a diversity of leisure and holiday opportunities;

quality public infrastructure; abundant food; effective medicine and healthcare – then there will be compromises, there will be trade-offs and there will be uncomfortable choices to make. The same applies to renewable energy. We need to accept this.

If we don't accept this, the outcome will be a reduction in living standards and life expectancy back to pre-industrial levels. I doubt whether many people would welcome such an outcome were it to become a reality.

If we do accept this, the outcome of a more sustainable future will be achieved. And it will involve a lot of mining.

7
Responsible and sustainable mining

We have considered what mining is, what it involves and why it's essential for sustainability and the energy transition. We will now contemplate the sort of mining, and the sort of mining industry, we want to have producing all those necessary metals and minerals.

Firstly, we will reflect on 'the sins of the past'.

It would be disingenuous of me to talk about mining, why it's essential and why it needs understanding and respect, without acknowledging that over the years there have been plenty of instances where the mining industry hasn't lived up to the standards expected of it. Sadly, there continue to be mining companies that don't pay sufficient regard to safety, environmental standards, sustainable operating practices, the well-being of communities in which they operate and financial transparency. And I'm not just talking about the companies with opaque ownership and little or no accountability that I mentioned previously. Regrettably, there are examples of poor behaviour across the spectrum, whether the companies responsible are publicly listed, privately owned or in government hands. It's important therefore to face up to, and learn from, the sins of the past, and to seek to eradicate the sins of the present.

How do we deal with mining companies that aren't behaving as they should? And why does it matter, if the majority of the industry is playing by the rules? Let's answer the second question first, as that will confirm the importance of working on an answer to the first question.

It matters because every industry, indeed every enterprise, should be working towards the highest standards in safety, environmental management, operations, community and finance. It matters because an industry that's so central to our sustainable future needs to be above reproach. The mining industry, so vital to the achievement of our sustainability goals, cannot have the focus on that achievement derailed by the practical consequences and valid public opprobrium that come with bad behaviour.

So, to address the first question, how do we deal with mining companies that aren't behaving as they should? The answer depends upon who owns them. As we have discussed previously, companies that are listed on respectable stock exchanges with comprehensive reporting, thorough auditing and the unrelenting attention of analysts and investors are much easier to influence for good than a mining company in private hands, or in the hands of an authoritarian state with little regard for transparency.

Mining companies listed on established stock exchanges, with institutional investors who care about these things, are at the forefront of responsible corporate citizenship. There are plenty of privately held companies that share this commitment to responsibility and transparency, and it's very important that we don't infer that private ownership per se equates to lower standards of corporate behaviour. Many privately owned mining companies can be rightly proud of their record on safety, sustainability, working conditions, community engagement, transparency and much else.

There are, however, plenty of mining companies which are listed on stock exchanges with lower standards, or not listed at all, and owned by individuals, entities or governments which don't have the same concern for sustainability, the environment, safe work practices, workforce welfare or anything other than making money or securing a supply of metal. Often this ownership is opaque, and it's difficult to know who the ultimate beneficial owners are, making it even harder to call poor corporate behaviour to account. But these exceptions shouldn't stop the responsible end of the mining community from seeking to maintain the highest possible standards, and also seeking to improve the public's understanding of what they do and how they do it.

So we should influence where we can, and the best way to ensure we can continue to influence for good is to encourage the investors in mines to keep them in the hands of companies that are subject to the level of scrutiny we describe above. And where mines are in the hands of state-owned enterprises of dodgy states, or in the hands of unaccountable private investors of uncertain provenance, we should still be doing all we can, including at the intergovernmental level, to hold them to account. Thinking that they will stick to sustainability agreements just because they say so is naïve in the extreme.

So on to some of the sins of the past (and the present).

Environmentally, miners have caused havoc, not just due to poorly designed, constructed and managed tailings facilities, but also with the indiscriminate use of some pretty toxic chemicals; the over-use and misuse of water resources; the fouling of rivers and groundwater; processing plants that pump all manner of noxious gases into the surrounding atmosphere; the badly planned and executed dumping of enormous volumes of mine waste; excessive and often unnecessary deforestation; and lots more.

An unfortunate example of this is happening in parts of Indonesia even as I write. Although there are some excellent examples of good environmental stewardship in that country, there are also some pretty awful examples of nickel mines and processing facilities which have been built, and are being operated, with little regard for the environmental consequences. Exacerbating this problem is the unfortunate fact that the mines and facilities in question are owned by overseas companies from countries with minimal (or no) transparency, which ship in their plant, equipment and construction materials, their workforce and even the workforce's food! So not only are they damaging the host country's otherwise pristine landscape, they are also contributing very little to the economic development of the communities within which they operate.

From a community relations perspective, miners have often treated the communities around mines, including the traditional owners of land they are exploiting, with Olympian levels of disrespect, neglect or both. The demolition, without consultation, of whole rural villages is not unknown, often with the acquiescence of corrupt government officials. Even when their actions haven't been that extreme, a lack of regard for the quality of life of those affected by a mine development has often led to some very bad outcomes. The diversion of rivers or other watercourses relied upon by local communities; sending an endless procession of lorries and heavy machinery through the middle of previously peaceful small towns; major construction without thought for those living close by; the pollution I mention above: these are just some of the negative impacts miners have foisted on local communities.

I was once asked if I would like to do some work for a mining company in Norilsk, back in the days when people travelling on British and Australian passports could happily come from and go to Russia. I said no, mostly because I couldn't countenance having to spend time in Norilsk, an area which has been denuded to the point of becoming a dystopian moonscape. I should note that I have happily spent time in Russia visiting gold mines hidden away in mountain regions where the environmental footprint was both managed and minimal, so I'm not taking a swipe at the whole Russian mining industry.

As I note elsewhere in this book, there have been plenty within the mining industry who have a fairly tenuous relationship with morality, and have been very happy to pay bribes, turn a blind eye to government corruption and even, in extreme cases, engage in (or pay others to engage in) extortion, threatening behaviour and violence. Mine owners have also been known to use the courts to cynically get their way, usually against individuals or companies which don't have the capacity or the funding to mount a credible response, particularly in countries where the rule of law isn't necessarily upheld. The Russians, back when their companies were active on western stock exchanges, were particularly known for this. To be fair, though, this is a phenomenon not restricted to the mining industry. Sadly, bribery, particularly in parts of Africa, is very much a 'sin of the present'. I discuss this further below.

Finally, mining has had a dismal safety record. Poor risk management; shoddy design and construction; a cavalier approach often encouraged by cost-cutting; little assessment of what caused accidents and thus little attempt at avoiding them in the future; a policy of blame-shifting, cover-ups and obfuscation when things went wrong: all of these things have contributed to a tragic number of deaths, injuries and near-misses that could easily have been even more deaths and serious injuries. Safety culture is something the industry has worked extremely hard to address over recent decades and with a great deal of success, but culture takes a long time to change.

There are plenty of other examples of poor behaviour, but I trust you get the picture.

In improving the standards of mine safety, operating practices, environmental performance, community relations and financial reporting, it's therefore important that responsible mining companies, and hopefully *all* mining companies, make it clear that poor behaviour has no place in today's mining industry. Happily, with the vast majority of miners, it doesn't. The fact that all responsible miners place safety, and their safety performance, front and centre of their reporting isn't just for show. It's to reinforce to their own employees, their shareholders and everyone they deal with that safety is their number one priority. It hasn't always been the case, but in almost all companies it is now.

With the industry's chequered history, changing perceptions require that miners' behaviour be not just better but above reproach.

Improving the image of mining means acknowledging the sins of the past and demonstrating through the way the industry as a whole operates today that those still sinning in the present are an ever-diminishing minority.

It's my hope that as people gain a better understanding of the mining industry, that ever-diminishing minority will find it within themselves to improve their own standards and join the cohort of responsible, sustainable miners.

~~~~~~~~~~

We discussed community contributions in chapter 4. Now let's consider in more detail mining companies' dealings with communities.

We have noted previously that many mines are built in remote locations where the mine is one of the few sources of gainful employment. There are also plenty of mines which, although not particularly remote, in the absence of much other industry, are still the main source of employment for the surrounding community. Generally speaking, these communities will probably welcome the arrival of a long-term mining project.

Conversely, sometimes a mine development will be proposed in an area where the community are relatively self-sufficient and don't want a huge mining project imposed upon their lives. In other circumstances, the construction of a mine may cause problems for existing industries, particularly agriculture, aquaculture and forestry. Tourism is often impacted by the arrival of a mine, particularly if the attraction to the tourists is remoteness, natural beauty and unspoilt wilderness. And then there are communities which exist specifically because of this wilderness and remoteness, and the arrival of a mine development takes away their whole reason for being there. Sometimes, although a community may be keen on the employment opportunities, they will have legitimate concerns about the environmental
~~~~~~~~~~

impact, be that air pollution, water pollution, vegetation destruction or even just visual damage.

Finally, but very importantly, traditional communities, known variously around the world as First Nations, indigenous or aboriginal communities, may claim rights over the land, or there may be elements within the landscape of cultural significance which would be threatened by the construction of a mine. There was a widely reported incident in Western Australia in recent times when a mine development team at Rio Tinto managed to blow up caves of enormous cultural significance to the indigenous community. The CEO ultimately lost his job as a consequence: a sensible outcome and a timely reminder never to forget the cultural element of working within a host community.

All this means that in planning for a mine development, and usually as part of the approval and licensing requirements of the relevant regulators and government authorities, mining companies need to spend quite a bit of time and energy dealing with the expectations, demands and needs of the communities affected. The concerns and well-being of communities are very important, very serious and usually very legitimate, but they can also be very time-consuming, very expensive and occasionally very frustrating. Thus they are a significant factor in the planning and development process for any new mining project.

By the way, I say the concerns of communities are 'usually' very legitimate, because there have sadly been situations where community engagement with mining developments has been hijacked by NGOs and charities with bigger agendas than just the well-being of the community in question, and locals have found themselves stuck in the middle of disputes on geopolitical, macro-environmental or ideological matters. This turns into rancour, unwanted court cases, delays sometimes of years, huge costs and unnecessary bad blood. Ultimately, the people who suffer the most harm from this are the communities, which were really only wanting to be consulted, be treated with respect, get some improvements to local infrastructure and services and benefit from enhanced employment opportunities.

Working constructively with communities involves a variety of factors.The first, of course, is genuine and timely engagement. This isn't as easy as it sounds, as the first thing a company needs to do is figure out who the community are, and who are the appropriate representatives of that community to do the engaging with. These representatives will probably include, among others, members of local government, business leaders and, in many countries, religious leaders. In countries where the local tribe is still the most important social grouping after the family, it will include tribal elders or leaders. In countries such as Canada, Australia and elsewhere, it will include elders and representatives from the area's First Nations/indigenous groups. Once you have figured out the appropriate people to consult with, you can, in a formal way, discuss the proposed mine development, seek the community's consent and support, find out what their main concerns are likely to be and discuss opportunities for local involvement, employment and contracting.

It's particularly important to ensure that, in dealing with traditional owners or tribal leaders, due respect is given to a connection to the land that's foreign to a lot of western cultures and thus to many of the people tasked with managing mine developments. The executive in charge of mine development for a London-, Sydney- or Toronto-listed mining company needs to make sure they understand, or are very open to learning about, the cultural significance that the traditional owners of the land attach to the site they are seeking to develop.

There's nothing like providing meaningful employment opportunities to get a community on side. We have noted that one of the things that makes mining projects hard to get off the ground is that they are necessarily so long term. When it comes to employment, however, this is a good thing. You may be offering the prospect of jobs at a mine with a life longer than the average worker's career span.

The other big opportunity for communities in this respect is the provision of goods and services to the mine. As we noted previously, mines need a lot of goods and services, and the more that can be sourced locally, the better. If a local community not only has workers directly employed by the mine, but also supplies goods, services and contractors, you are ensuring that the economic benefits of the project are shared and genuine goodwill is generated.

If the mine is big enough, it may even be that company-sponsored training can be established, to ensure the next wave of mineworkers develop basic skills and knowledge, ready to work in the mine down the road or up the hill from their homes.

Another significant area requiring community consultation is management of the environment and protection of local amenities. If you are going to have an enormous open-pit mine a mile from your front door for at least the next fifty years, you want to know that your quality of life will go up, not down. That means the company developing the project must satisfy the community as to noise, air pollution (including dust), smell, and, for that matter, visual changes. If you are used to living next to a tree-covered mountain, and that mountain is to be gradually dug away until it turns into an enormous hole in the ground, you probably need quite a bit of consulting and reassuring!

The aspect of community that gets most airtime is probably the provision of infrastructure, facilities and services. This is the idea that a mining company, as part of their 'social licence to operate', should benefit the community through the provision of facilities for the public good such as schools, hospitals, clinics and sports grounds. In this respect, it is very important that those running the project consult thoroughly. I related the story in chapter 4 about the unnecessary school provided in one African country where what the community really wanted was a decent bridge across the river to the existing school. The message coming from this cautionary tale is: consult thoroughly, and consult with the right people.

Assuming the company looking to develop a mine do consult thoroughly, and with the right people, this should be a very positive experience for both the miner and the community. And it makes good sense. In return for what's effectively a massive and long-lasting invasion of the community's privacy, or, more to the point, a complete and irreversible change to the status quo, the community will receive tangible, long-life public assets for the benefit of them all.

Unfortunately, it has become common for host governments to see the arrival of a mining project as an opportunity to effectively outsource the provision of public services to the incoming mining company. As part of the licensing requirements for the mine development, they can get greedy. The mining companies have, for the most part, responded well, building roads, schools, hospitals, clinics and sports fields, constructing water supplies and reticulation, power generation, indeed just about all public infrastructure in some cases. This is wonderful for the communities in terms of access to services. But it also means they sometimes come to see the mining company as a proxy for the government, and thus blame them if they don't get the level of services they expected.

For example, a common problem will be that the miner will build, as required, a school or a hospital, but the government education authority or health authority will then not provide enough teachers, doctors, nurses or equipment to actually run them properly. Weirdly, the miner who built the physical structures will then get the blame when the school or hospital can't operate properly due to a lack of staff and resources from the education department or the health department. Roads and sports facilities built by miners and then not maintained by local authorities are another very visible source of frustration.

In a worst case but nonetheless unfortunately realistic scenario, when you live in a dysfunctional community with no money, and you walk along the pot-holed road, past the empty school, the under-resourced hospital and the unmown, weed-infested football pitch, all with signs noting the mining company that built them, it isn't surprising if the aforementioned goodwill starts to evaporate, even if these failings aren't actually the company's responsibility.

In response to government's inability, or disinclination, to properly resource and maintain the infrastructure built by miners, the mining companies have usually stepped in and done it for them, going well above and beyond the licence requirements. This is to the great credit of the mining companies, but it further reinforces the impression that they have become the de facto government, at least as far as public services go.

The last 'front-end' community consideration is upgrading existing infrastructure to cope with the increased capacity requirements of the mine itself, such as water, power, roads and railways. Again, this requires careful consultation and agreement. Driving a new railway line through the middle of ancestral or tribal land; proposing to cut a village in half by upgrading the single-lane main street into a road capable of carrying haul trucks; planning enormous electricity pylons across the community's mountain view: all will require the acquiescence of the community, and it shouldn't be assumed that that acquiescence will be forthcoming. Or, at least, it may not be agreed in the form that the mining company proposed it. The upside for the community may well be improved water supply (or, indeed, *a* water supply), more reliable electricity, higher-capacity broadband and better all-weather roads. But even so, this will be yet another area to be consulted upon, negotiated over, hopefully finally agreed upon and budgeted for.

Finally, it's vital, right from the start, to give consideration to what will happen after the closure of the mine, regardless of how far away that might be. After all, if the mine employs half the town, who then spend their money in the shops and on the services run by the other half, there will be a very significant adjustment to be made when the ore runs out and the mine salaries come to an end. This is a huge issue right now in many parts of the world, and perhaps particularly in places like South Africa, with its very rich mining history, where many mines are coming to the end of their economic life. So planning for this post-mine life is really important. Can agriculture take the place of the mine? Can the processing plant structures be repurposed for manufacturing of some kind? Is there a realistic prospect of future brownfield or greenfield exploration resulting in new deposits to extend the mine for another generation or more? All this needs to be contemplated right at the beginning, for the sake of both the mining company and the community.

In summary, miners need to spend time, energy and money; be respectful; consult carefully and with the right people; be prepared to amend plans where there is disagreement; be ready to step in when governments disappoint; and think for the long term. If they do these things, there is every reason to expect a welcoming local community and a willing local workforce who will support the mine even when not everybody does.

~~~~~~~~~~

In discussing the challenges of dealing with communities, I was being (justifiably) rude about some local and provincial governments and their habit of outsourcing their responsibilities to mining companies. Let's now consider in broader terms the influence of governments on mining. Among all the uncertainties of a new mining project, there will always be one certainty, and that's the influence of government. And, generally speaking, it's nice to have that certainty. Trying to operate mines in regions where there's civil war, or a military coup, or smaller-scale armed insurrections, or just plain old-fashioned anarchy is difficult, and I'm sure makes the companies operating there dream of the certainty of even the most interfering, bureaucratic and regulation-loving but stable administration.

Government influence starts right at the beginning. In many jurisdictions, the responsible government department will have carried out a geological survey and will have
~~~~~~~~~~

a fairly good understanding of the main mining regions and what might be there. Places as diverse as Russia and Australia have very detailed government-held geological data, available to anybody for a fee, and that's the logical place to start when thinking about exploration.

Once you have studied available geological data or performed initial geological work of your own, and have formed the view that there's a potential deposit in a particular location, you will need either to make a claim over the location (referred to in mining parlance as a tenement) or to apply for an exploration licence, or both, depending upon the rules applicable in the jurisdiction.

It's once you have decided that you have a deposit worth developing that the interactions with government really start in earnest. Remember that at the same time you will probably be trying to sort out financing, and often you won't be able to secure that until you have the permits, but you won't need any permits if you can't get financing, and so it can all get a bit chicken and egg. Anyway, the first thing you will need is a development permit, and then you will probably need permits for things like water usage and access road construction. Then you are likely to need specific construction permits and agreements once you start actually building things. You are still a long way from actually digging any holes in the ground and you will already be drowning in government-issued permits.

At the same time, you will need to sort out the environmental side of things and it's almost certain that you will need to do an environmental impact assessment (EIA). These can take a very long time, as in, years. The ease with which you are able to do your EIA, ending up with an EIS (environmental impact statement), will depend to an extent upon the attitude of the government department you are dealing with. If they start from the position that mines are bad and environmentally destructive, then you are in for a long and difficult journey, potentially only getting there in the end if a political decision is made that the government of the day wants your mine to go ahead. Even if the relevant department takes a more measured and practical approach, it will almost certainly take more than a year, probably two, and will also be quite expensive. Environmental consulting costs can really add up. This isn't in any way to belittle the EIA process. It's very important to ensure that the mine's environmental footprint will be minimal and that any particular environmental issues can be dealt with appropriately.

In many countries, you will also need to deal with native title, so formal agreements with traditional owners or First Nations people will have to be negotiated and agreed in accordance with legislation. This is vital, but again it can take a long time and will probably involve a lot of listening, probably some compromising, a lot of consultation and a lot of consultants.

The next area where you will need government support and agreement is in the provision of infrastructure. Your mine will need road access, power, water, perhaps rail if your mine development is for a bulk commodity, and all these things will require you to negotiate with a government authority to agree terms for the provision of a connection to your project. Of course, mostly the negotiation will come down to how much you are going to pay. Obviously within the mine gate, the cost is entirely yours. But if you are the only customer at the end of a specifically built 160-kilometre transmission line to get to your mine gate, then it's reasonable that you should be contributing towards the cost of that. How much you contribute is yet another matter for negotiation and, potentially, arbitration. It all gets more complicated if there's an expectation that over time there will be additional customers taking power from the transmission line that the government would like you to pay for. So these things are never straightforward. Roads, railway lines and water pipelines are all similar. To what extent should you contribute towards infrastructure projects which, although you might be the only customer at the moment, are nonetheless public works with the possibility or probability of other customers in the future?

A major area of government influence, as we have discussed in chapter 4, is tax. If a government really wants you to develop a mine, perhaps because of significant local employment opportunities, then they may be keen to negotiate favourable tax breaks or reduced royalties. If they don't care about your mine, or, as a consequence of their political leanings or long memories for historical failings by the industry, they are of the view that miners are exploitative, then you can expect they will be looking for all they can get from you. Add that to the list of potentially fraught and drawn-out negotiations. Fortunately, many governments, including in the developing world, who have started their tenure with a very populist anti-mining approach, have softened that approach once faced with the reality in government of having to pay the bills.

Once your mine is commissioned and in production, there will be continued government influence through renewals of mining licences, export licences, infrastructure charges, environmental monitoring, the country's employment laws, ongoing taxation and royalties and no doubt much else.

Finally, governments are, quite rightly, concerned to ensure that mining companies do the right thing in relation to preparing for the end of the mine's life. Rehabilitation and potential alternative uses for the mine property, consideration of alternative employment opportunities for the mine workforce and assessment of the broader impacts on the surrounding communities are all things which governments will probably require agreements on. There may also be a requirement for the lodging of bonds in relation to rehabilitation and other post-closure costs. More assessments, more consultants, more negotiation, more costs.

All of the government interactions I have mentioned above are reasonable and necessary to ensure a sensible, well-planned and sustainable mining industry which will benefit both the miner and broader society. I say this on the assumption that government ministers, politicians and civil servants are approaching the mining industry in good faith and as something to be encouraged for what it is: a responsible and sustainable provider of the resources the world needs for life and for a sustainable future.

~~~~~~~~~~~

One of the essential skills required in any business, and perhaps particularly so in mining, is the ability to understand and deal with political considerations. A large mining project impacts a lot of people directly and indirectly; it impacts the environment; it has a significant economic impact on a host country or region; its products may be of strategic importance; it provokes a lot of strong feelings from all sides of the spectrum; and it's going to be there for a long time. So it's always going to be the subject of considerable political interest.

For all these reasons, miners need to approach their dealings with governments, politicians and public officials with their eyes open. This doesn't mean assuming either a lack of trust or any intention to make life difficult. It means understanding that when mining companies are dealing with governments and government agencies, they need to appreciate that the latter will probably have agendas which go beyond the simple objective of assisting in the development of the mining project in question.

These agendas can be both negative and positive. One negative example is governments burnishing their green credentials by being seen as tough on polluting industries, something they achieve by making broad generalizations as if all mines automatically fall into this category. A particularly damaging one is invoking resource nationalism, where they promote the notion that all the profits from the mining industry should stay in the country. They usually choose not to mention, or conveniently ignore the fact that, firstly, in many cases, there would be no serious mining industry without foreign investment; secondly, the country benefits by receiving significant profits from the mining industry through the taxes and royalties paid by miners; and, thirdly, the country also benefits from the multiplier
~~~~~~~~~~~

effect of mining wages and money paid to local suppliers of goods and services. A final agenda can be using the mining industry as a convenient focus of blame for the sins of the past, be that the behaviour of miners from another age, the environmental damage caused by unrelated industries or someone else's corrupt practices.

At the extreme end of the resource nationalism I mention above is the threat of the nationalization of mines, where governments expropriate the assets of mining companies and take them into state ownership, often without proper compensation (and sometimes with no compensation, or compensation so derisory it's meaningless). Unfortunately, this happens, and when it does, the struggle, usually through the courts, to either get the mine back, or at least receive appropriate compensation, can drag on for years or indeed for ever. Something to think about.

On the positive side, governments' agendas can include promoting mining as a way to generate economic growth and employment, both directly through the mine and indirectly through the provision of goods and services, demonstrating self-sufficiency in certain minerals, particularly 'strategic metals' such as gold, levelling up the balance of trade and encouraging foreign investment.

So the attitude of a host government, and therefore whether its influence on mining projects will be on balance negative or positive, may depend upon their underlying agendas. Hopefully, though, whatever the agendas, there will be some thought put into their approach to the mining sector and they won't make short-term decisions so as to be seen to be doing something. There's a real risk, in the current environment, of this happening, particularly in relation to environmental matters.

Finally, the unfortunate reality is that not all governments operate with the best interests of their people, local businesses or inbound investors at heart. And mining companies can and do get caught up in this. It's very difficult to operate in the face of government hostility, the absence of the rule of law, rampant corruption or a sense that promises, undertakings and contracts may not be honoured. But, remarkably, mining continues even in these circumstances.

As well as dealing with the sometimes unpredictable manoeuvrings of those in power, miners must also cope with issues caused by the political machinations of individual politicians, political parties, trade unions, NGOs, lobbyists and self-appointed opinion leaders, most of whom don't really know what they are talking about when it comes to mining. And, of course, miners themselves have agendas, the first one usually being to make a profit. It would be reasonable to say that just about everyone has an agenda. Miners need to ensure they understand what they are, and seek to balance the competing agendas (including their own) to ensure the best outcome, hopefully, for everyone.

We have previously dealt with all the areas in which governments impact upon the practical realities of developing and running mines, including regulation, taxation, licensing, underlying attitudes and all the rest. But the rogues gallery of political interests I note above bring a whole lot of less practical but often even more difficult-to-navigate issues, and I provide some examples below.

Recently, a proposal was floated to develop a metallurgical coal mine in regional England, primarily to provide coal for the domestic steel industry. There was, and continues to be, much uninformed fury at the notion that the UK government would even dream of sanctioning 'a new coal mine'. Of course, the people carrying on about this have no idea of the distinction between thermal coal and metallurgical coal and blithely assume that all coal is used for burning to generate electricity. The public discourse wasn't helped when a very senior politician, indeed at the time the country's most senior politician, expressed the view that it would be madness for the government to approve 'a new coal mine'.

What those objecting to the project didn't, or wouldn't, understand was that the project in question was for the purpose of supplying metallurgical coal required for the steel-making process, replacing coal which is currently shipped from the United States. In other words, the primary purpose of the mine was to satisfy local demand for metallurgical coal which is currently being met from overseas. Development of the mine would significantly reduce emissions by removing the need for all the UK's steel-making coal requirements to be carried across the Atlantic Ocean in bulk carriers, while the emissions from the steel mills would remain exactly the same. It would also improve the UK's self-sufficiency, contribute to the security of its supply chains, be more flexible in the capacity to supply customers, provide employment, provide orders for contractors and machinery manufacturers and be a significant gain for the country's balance of payments.

Initially, the mine was approved, although, following a change of government, it has just been unapproved. So, sadly, this is a very current example of another problem miners face: political opposition from those wishing to advertise their 'green' credentials without considering all the implications, without understanding the underlying necessity for essential metals and minerals and without realizing that if they would only look at this issue holistically, they would realize they are contributing to a less sustainable future.

Political point-scoring has been a problem for miners for a long time.

A second example: Australia has probably the most abundant resource of high-quality coal, indeed without question the cleanest thermal coal, in the world. Activist and political pressure is being brought to bear to reduce and ultimately end the extraction of thermal coal in Australia. The Australian public is very environmentally aware and thus the push to end thermal coal mining has strong support in many sectors of the community. However, just like my previous example, the simple answer 'just stop coal mining' would actually be environmentally counterproductive.

The process of mining coal isn't particularly environmentally sensitive. You don't have to do any processing of coal, so there's no massive energy or water usage involved in a coal mine. As it isn't hard-rock mining, the blowing-up part is also not as energy intensive. So digging it up and shipping it off to the nearest port is fairly straightforward and certainly not environmentally damaging, other than leaving some quite large holes in the ground in places that very few people visit. The environmental impact from thermal coal is in the burning of it in power stations to produce electricity.

And this is the bit that no-one seems to appreciate. The burning of thermal coal to generate electricity isn't going to stop in the short to medium term, as noted in chapter 4. The countries which are continuing to generate most of their electricity from thermal coal aren't going to be rushed into phasing it out because western virtue signallers want them to. What they want is abundant, cheap, reliable electricity, just like the West has had for almost two hundred years. So, as noted previously, they continue to burn coal and indeed are busy building many new power stations so they can burn even more. The short- to medium-term burning of significant volumes of thermal coal is a given.

So you can either sell these electricity generation companies relatively clean, very efficient Australian thermal coal and thus minimize the environmental impact, or you can condemn them to burn much more polluting, dirtier, lower calorific-value coal from somewhere else, like Indonesia, India or China. Unfortunately, the direction of travel at the moment is towards the latter. If this happens, we, as a global community, will all be far worse off. All the virtue signalling in the world won't change the reality that forcing utility companies to use dirtier coal than they need to is going to lead to a far worse environmental outcome. As the atmosphere doesn't have walls, burning inferior coal in one country is bad for all of us.

In the longer term, we would all like to see an end to the burning of coal. But in the interim, let's at least ensure we are burning the cleanest coal we can get our hands on. Ensuring a sustainable future includes making sure the pathway to that future is as sustainable as possible. This potential outcome for Australian thermal coal isn't contributing to a sustainable future; it's naïve, blind ideology in action.

A third example of the difficulties miners face in dealing with politics is the impact of the admittedly much broader issue of nimbyism. This is a particular problem in western countries. A current example is a copper project in Arizona that has faced almost interminable delays due to objections from all sorts of interest groups. Arizona is a very large state, has a rich mining heritage, is sparsely populated outside a few major cities, is mostly of unexceptional landscape and in all respects should be the ideal place to develop a mine. It's particularly unfortunate that these objections are thwarting an enterprise that would enable the United States to produce a much large portion of its copper requirements domestically, reducing its reliance on imported metal. It's estimated that this mine, by itself, could supply 25 per cent of the US's copper needs.

It's the same old story: you can produce essential metals domestically, to high safety, environmental and quality standards, or you can import metal from somewhere else, with all the uncertainties, and added emissions, that brings. It would appear that for many people in the United States (and plenty of other countries), they say they want the former, but only on the proviso it doesn't happen anywhere near them.

Thus we have nimbyism, dressed up as environmental responsibility. Opposition comes from people and groups who want all the benefits of a developed world lifestyle that this metal enables, but without a mine that actually produces it. This is particularly odd for a society which is already totally dependent upon electricity, and becoming ever more so. I'm not sure how they expect to have that electricity without the copper that enables its transmission.

What's needed, once again, is strong political leadership from appropriately informed representatives who can cut through the misinformation and ignorance, and enable projects such as this one to proceed, bringing with it all the benefits and certainties of a strong domestic supply of critical minerals.

This brings me to my final, and perhaps the most significant, example of uninformed political posturing leading to exactly the opposite outcome to that which was naïvely hoped for: the issue of critical minerals which we discussed in more detail in chapter 6. Short-term decision making, usually to appease whichever segment of the voting population is being focused on at the time, has led to either poor policy or, more commonly, simply no policy in respect to critical minerals. Difficult decisions involving trade-offs, for example accepting some level of domestic 'dirty' industry in return for energy and resource security, are just not made. And consequently the UK and similar countries are left overly reliant upon overseas suppliers for almost all their critical minerals needs. This has been a win for the nimby brigade and those who won't accept any level of compromise in order to develop mines and mineral processing plants, but a potentially dangerous loss for the countries' energy and resource security. What the climate absolutists don't seem to appreciate is that in the medium to long term this may well turn into a significant loss for our way of life and our collective well-being. The really terrifying thing is that perhaps the climate absolutists don't care. Perhaps they *would* care if they understood that the overseas suppliers of critical minerals usually have little interest in the environmental purity they hold so dear. As I have noted a couple of times already, the global environment doesn't have walls.

On a positive note, the development of critical minerals strategies in the UK, the United States and other jurisdictions is a step in the right direction and hopefully meaningful domestic critical minerals industries will follow.

It's very frustrating for the mining sector to know that the opportunity to provide the minerals and metals necessary for a sustainable, reliable, secure future is there, if only we could find politicians with that rare combination of brains, backbone and even slightly longer-term thinking to help make it happen.

~~~~~~~~~~

Now to a trio of issues which contain a lot of negatives but need to be addressed: supply chain integrity; the related but broader problem of corruption; and illegal mining.

An issue which rightly receives a lot of attention in the broader world of commerce is supply chain integrity. For those unfamiliar with the term, it refers to ensuring that not just your own business, but the supply chain for your business, is upholding appropriate ethical and moral standards in relation to its workforce, sustainability, corruption, governance and business practices. Mining has a part to play in this.

Common areas of concern when discussing supply chain integrity in the context of mining include, among other things, modern slavery, child labour, environmental practices, conflict minerals and the all-encompassing issue of corruption.

The principle behind all this is that a company have a responsibility to ensure they aren't indirectly contributing to human suffering, environmental degradation or corruption by purchasing materials, equipment or services from companies or organizations that engage in these practices. The further principle is that it isn't enough for a company to claim ignorance, but that they should make reasonable efforts to ensure the integrity of their suppliers and their supply chain.

The example that has received a lot of headlines in recent years is the use of child labour to mine cobalt in the DRC. The major mining companies and the major car manufacturers have taken steps to ensure that none of the cobalt they sell, trade or buy for use in the manufacture of
~~~~~~~~~~

EV batteries has been mined using children, and this is a very good thing. They are given encouragement to do this by legislation such as the Modern Slavery Act in the UK and similar laws elsewhere in the world.

Unfortunately, there are plenty of companies and traders who don't care about these things, and certainly don't care about a piece of British legislation, and there's plenty of evidence that they are operating in the DRC, 'aggregating' cobalt supplies from undisclosed sources and then selling them on to traders and buyers operating for private or state-owned companies in countries that are happy to turn a blind eye to their questionable provenance. And thus we come back to the same old problem I have flagged up before. It's vital that reputable manufacturers, miners and everyone else in the supply chain seek to ensure the provenance of goods and services. But when there are supply chains that, from raw materials through to finished goods, are owned by those who don't care about these things, and they operate in jurisdictions that allow them to get away with it, then there's very little that can be done.

Or is there? Consumers can be alert to situations where there's a high chance that the product they are considering buying has, somewhere in the supply chain, involved modern slavery, child labour or negligent environmental practices. And then they could decide not to buy that product. Unfortunately, for a lot of people the fact that the item in question is very cheap often means that in the battle between wallet and conscience, the wallet wins. The fact that the product was ludicrously cheap is both the indicator of the problem and the reason they succumb and buy it anyway. More effective would be for responsible governments to be alert to such practices and, where there's clear evidence to back them up, ban the import and sale of these products. Such decisive action doesn't seem likely.

Another concern for supply chain integrity is the issue of conflict minerals: that is, minerals that have been mined in areas of armed conflict, and are used to fund the ongoing fighting. You may not be surprised to know that a big problem country for conflict minerals is the DRC, but it's by no means the only one. Mining companies, together with reputable consumers of these minerals, are using technology such as blockchain to give greater assurance as to the provenance of metals in their supply chains. Diamond miners and traders have long used the Kimberley Process certification scheme to enable the identification of individual stones to confirm their origin.

The final and in some senses overarching issue in relation to supply chain integrity is corruption, but as that's wider than just the supply chain, I discuss it in detail below.

So what's the extent of the miners' responsibility in relation to supply chain integrity?

I would suggest it includes doing the following: maintain the highest standards in their commercial dealings and their own governance; participate in global efforts to monitor and where possible eliminate modern slavery and child labour in supply chains; stand firm on corruption wherever it rears its head; and work with other miners, governments and international trade and industry bodies to develop effective methods for identifying the provenance of metals. These are some of the things miners can do.

It's not the job of the mining industry to be the world's commodity police force (although I am sure some governments would love the idea of outsourcing policing to mining companies in the same way they seek to outsource the provision of community assets!). But I do believe miners, particularly those who operate internationally and with global supply chains, have an obligation to assist in the elimination of those practices, sadly still prevalent in parts of the globe, which destroy lives, damage the environment and make the task of mining for essential minerals that much harder.

~~~~~~~~~~

The second negative issue facing miners is corruption.

A number of years ago, I visited a gold mine in Egypt. To get there, I flew from London to Cairo, caught a
~~~~~~~~~~

connecting flight to Hurghada by the Red Sea, followed by a four-hour drive down a dusty highway to the mine. As Egypt had not long before gone through a military coup, which not everyone was happy about, there were roadblocks at fairly regular intervals along the highway as part of the new government's efforts to curb unrest. At most of these checkpoints, we were waved through, but on one occasion we were pulled over.

This particular roadblock was run by an imposing, solidly built man in civilian clothing with a military cap and a sizeable revolver on his hip. In supporting roles were about ten very young-looking soldiers with rifles, which they pointed vaguely at our car. They looked not just young but nervous, if not terrified. I wasn't particularly worried about them intentionally shooting us. We didn't, after all, look like supporters of the previous government, the people whose progress down the highway the roadblock was primarily seeking to hinder. No, I was much more worried that they might shoot us accidentally. They looked so nervous that I remember thinking it might only take one ancient lorry backfiring as it trundled past the checkpoint for them to squeeze their triggers in fright.

Anyway, we were asked to get out of the car, which we did. After looking at us for a minute or two, the man with the big revolver asked us to produce our passports, which we also did. He then wanted to look in the car, which our driver patiently allowed. He then wanted to search our luggage. At this point, our driver obviously decided enough was enough. He protested, in Arabic, and then made a fairly definitive statement, after which the main man expressed some irritation but indicated that we should get back in our car and go on our way. So we did that, quickly but with no sudden movements, happy that none of the nervous boy-soldiers with big guns had inadvertently let them off in our direction.

Once safely on the road again, we asked the driver, who spoke excellent English as well as Arabic, what he had said that had elicited such an effective response. He said, and I quote: 'I told him you were British businessmen and you wouldn't pay bribes so there was no point trying.' A very useful international reputation to have. And particularly useful when you spend time visiting countries where there's a fair bit of trying.

This anecdote segues nicely into a discussion on corruption. Clearly, the man at the checkpoint wasn't looking for miners in particular to help augment his military salary. But he highlighted the problems facing those working in and with the industry in dealing with corruption in many of the countries where miners operate.

And so we come to yet another issue to be navigated by the mining sector: how to manage the intersection of government influence and the sometimes questionable ethics of government employees and those with some degree of authority. This works both ways, of course: there would be very little bribery and corruption if there was a blanket refusal on the part of the mining industry to participate. Unfortunately, due to the massive opportunities often seen to be available, there will always be those prepared to run the risk and allow greed to get the better of them.

Corruption takes many forms. It can be as simple as a government official asking for a bribe in order to approve an exploration permit; it could be government employees looking for favours in order to facilitate a mining licence, or make favourable planning decisions around infrastructure. It could be a judge or some other official in the judicial system looking for an inducement to make a favourable decision. It could just be payments to officials to make things happen faster. In Bangladesh, I discovered, this even has a name: *speed money*. You want that environmental approval within the next six months? Just pay the standard, regulated fee, and it will probably happen, with a bit of luck and if the departmental staff are feeling generous. Pay some speed money and you can have it next week. The difference between speed money and the fairly normal practice of being given the opportunity to pay a higher fee for faster service (or 'express service') is that the fee for faster service goes to the institution or the government agency. Speed money goes into the pocket of the individual, and is thus a bribe, whatever name you give it.

Another example I have had the misfortune to come across was a company operating in Nigeria paying for a remarkable number of government energy ministry staff to attend a conference in Singapore. In return, permits got processed with impressive efficiency once they were back from their 'conference'. A final example involved providoring, the supply of food, alcohol and provisions for ships, including all those bulk carriers transporting commodities around the world. It was common practice that in order to obtain a contract to supply ships visiting certain ports, the providores would need to provide inducements to the captain or the purser or whoever onboard the ship was in charge of purchasing. The inducements seemed to vary depending upon the nationality of the ships' officers: for some, it was simply cash in a brown envelope; for others, it was fine Scotch whisky not otherwise readily available to them; and for others it was prostitutes. The practice had sadly become so normalized that the cost of providing these inducements was an allowable tax deduction in some countries, including, of all places, Australia.

The examples of corruption above are fairly straightforward, and avoiding them is also straightforward, as long as you maintain a strong moral compass. The implementation of legislation such as the US Foreign Corrupt Practices Act and the UK Bribery Act has also helped, by providing legislative clarity for US and UK companies, including making it clear that they could be prosecuted for engaging in bribery and corruption. This is helpful 'backbone' if the needle on your moral compass might otherwise start wobbling.

Where it gets messier is when the lines are blurred, where one person's bribery is another person's commercial transaction. There's also the significant issue of middlemen. When you pay what seems like a reasonable 'consulting fee' to someone in the DRC or the Central African Republic or, for that matter, Argentina or Pakistan (not wanting to pick on Africa all the time) to assist you with your mining licence application, is it your problem if some of that fee ends up in the hands of a government official or even a government minister? And should it be of interest to the payer that although that fee might seem reasonable in the context of the sorts of rates paid by western mining companies to western consulting firms, perhaps it's actually an extraordinarily large amount relative to the salary of the official or minister in question? Facilitation fees, dubious consulting fees, additional charges for unverified services, speed money. . . Do you look the other way on the basis that the additional costs are negligible in comparison to the revenue your expedited mine will generate? Do you convince yourself that this is just part of the 'cost of doing business'? Or do you get a new moral compass?

Corruption goes both ways in terms of the money flows. Handing state contracts to friends in return for political support and kickbacks is a big industry. A classic, non-mining example was the price per kilometre paid for roads being built in advance of Russia hosting the Winter Olympics. The only explanation, without mentioning the 'C word', is that those roads must have been quite literally paved with gold.

For those who are naïve or selfish enough to consider that corruption is a victimless crime, or simply the equivalent of an unofficial parallel economy, I urge them to look at the impact on countries like South Africa. The African continent's biggest and in many ways most developed economy has been brought constantly to its knees by endemic corruption. You only have to look at the consequences for their state airline (now bankrupt), their state electricity company (unable to generate sufficient power), their state railways (falling apart) and many other bodies to see the impact. And that impact is felt hardest by ordinary South Africans. Many of those who benefited are now sitting safely with their ill-gotten gains in Dubai.

Any government which engages in the practice of inflated contracts and subsequent kickbacks is diverting public money into private hands, and that public money is then unavailable to be used for the public good. Corruption is absolutely not a victimless crime.

It's very important to say, though, that navigating the spectre of corruption can be very difficult. Refusing to participate in dodgy contracts; declining the offer to provide facilitation fees; refusing to pay blatantly demanded bribes: all of this can make operating in some environments very difficult and indeed dangerous. Starting with delays, uncooperative bureaucrats and direct refusals to deal with requests; through to seizure of mine properties and deportation of essential staff; and all the way up to intimidation, threats of violence, actual violence and murder: these are all potential consequences of taking the moral high ground, and a miner with that fully operational moral compass needs to be ready for them.

So how *do* mining companies navigate such environments?

Fruit flourishes in the sunlight, and miners seeking to navigate their way ethically through corruption-infested waters can help their cause by letting in as much sunlight as possible. Transparency in your business dealings and contracts; being very clear about what you are doing; regular and comprehensive reporting, particularly in respect of related party transactions and payments to governments; clear statements of intent around ethical behaviour: all these things will assist in the fight against corruption. If officials know that their dubious behaviour may be disclosed, they may be more reluctant to engage in it. Where companies make very public and very clear stands against corruption, host governments are likely to make more of an effort to weed out, or at least discourage, the corrupt officials in their ranks.

Extra-territorial legislation such as I noted earlier is obviously a great help. So, too, is a united front against participating in bribery and corruption. Weeds are much easier to root out when they are small, and stopping bad practices while they are still small-scale is a lot easier if the industry as a whole is clear on its opposition to them. Once again, though, it's easy to say this, and to do it, with listed or private companies with strong governance, inquisitive investors and active regulators. It's a lot harder to do it with companies with opaque or unknown ownership held through jurisdictions which enable them to operate away from prying eyes. The fight against corruption will be a long one, but its effects are insidious and invariably it's those who can least afford the impact who are impacted the most. So the fight is an important one. Corruption and sustainability don't go together.

~~~~~~~~~~

The third in my trio of negative issues is illegal mining.

I once visited an underground gold mine in South Africa that was struggling with the problem of illegal mining. The mine had good grades, but no visible gold, so I'm not sure how successful the illegal miners would have been. The biggest problem was that they worked in gangs, and, of course, what happens when you have gangs is they start fighting with each other over territory. So rival gangs of illegal miners started having very violent and sometimes deadly fights over who was going to attempt to break in to the mine and steal potentially gold-bearing ore. The outcome of all this was that the mine security team, when doing their patrols of the perimeter fence, would sometimes come across dead and injured would-be illegal miners, the fall-out from the latest territorial conflict. Horrible for the security guards to have to deal with, a significant problem for the mining company and often pretty terminal for the gang members.

So what do you do about it?

A good start might be for those outside the industry to stop calling it 'artisanal mining'. It isn't, generally speaking, the work of artisans. It's the work of criminals breaking into legally acquired mine properties and attempting to
~~~~~~~~~~

steal from legally developed mines. The societal factors and the desperation that may have driven these people to criminal activity is a major issue that needs to be addressed, a bit like the early convicts sent to Australia from England for stealing a sheep or a loaf of bread. But it's illegal all the same, and giving it a euphemistic name only downplays the seriousness of what's going on. It's also very disruptive for the communities around the mines, who supply the legitimate labour force which does the legal mining, providing for themselves and their families and spending their wages with local businesses.

Making the situation even worse is that illegal mining is often carried out by children, or by both children and adults forced to do so against their will. We have heard about the problem in places like the DRC and its cobalt mines, but it's far more widespread than that, as we noted above when discussing supply chain integrity. For these people of all ages, the outcomes can be disastrous, indeed fatal. A recent tragic example involved a mine in the DRC which wasn't operating due to, among other things, an unstable pit wall. Despite barriers and other security measures, large groups of illegal miners took advantage of the idle mine and entered the property. Many were subsequently crushed by the inevitable collapse of the unstable pit wall that had led to the suspension of mining in the first place. The subsequent sensationalist reporting named the company which owned the mine, and a quick superficial read of the press reports could have led to the impression that this was somehow their fault. But it wasn't, it was the fault of illegal miners who ignored clear warnings, broke through barriers and trespassed on a dangerous mine property, with predictable and tragic results.

There are, of course, bigger issues at play here. Endemic poverty leading to desperation; a sense of injustice felt by those who, perhaps unlike their neighbours, don't have stable employment in the mine up the hill; anger at the government who may not be providing the basic level of services they would expect, helping them justify their 'right' to attempt illegal mining: these can all lead to actions which in a more stable environment may not take place. There's no illegal mining to speak of around the Australian or North American gold mines, for example.

As noted in our earlier discussion on communities, the provision of many community assets by mining companies as a condition of their mining licences may be inadvertently contributing to the problem. Where governments have essentially outsourced public services to mining companies, the miners can come to be seen as proxies for the government, and thus the focus of community grievance, which again, in some people's minds, helps legitimize illegal mining. Perhaps the solution is for mining licence conditions to no longer include the provision of community assets, but rather to include a slightly higher royalty, or some other indirect mechanism to provide host governments with the funds to then develop and maintain mining community public services on their own account. Unfortunately, history tells us that this may work in some places, but in others the host government will pocket the miners' cash for use on pet projects elsewhere (like their Middle Eastern boltholes) and the hospitals just won't get built. No doubt the mining company will then get the blame.

Illegal mining is unfortunately a problem for which there's no easy solution. It is, however, an issue that central governments, local authorities, charities, NGOs and the mining companies themselves need to address in a firm but compassionate way. The response needs to be firm because the consequences of unfettered illegal mining are often fatal, and it's also not fair on those workers from the communities surrounding mine properties who have genuine mining jobs. But it also needs to be compassionate because of the many underlying issues which can lead desperate people to try illegal mining in the first place.

Supply chain integrity, corruption and illegal mining are serious issues which are important to flag up in a book on the essential nature of mining because constant vigilance is necessary if the mining industry is to responsibly and sustainably provide the metals and minerals the world needs.

~~~~~~~~~~

The final issue to discuss in this chapter isn't about mining at all, but, rather, the lessons the mining industry can learn from the oil and gas industry. Its travails provide some very pertinent lessons and warnings, particularly as we consider what constitutes responsible, sustainable mining.

The oil and gas industry has suffered from a lack of understanding of how reliant the world is on what it produces. It has suffered from a poor image. It has had many spectacular, and very serious, calamities (think of oil rig and tanker fires, think of oil spills and the consequent environmental disasters). It has suffered from investor and financing rejection. It has also suffered from having players in the industry who haven't upheld high standards with respect to safety, the environment and governance.

With regard to a lack of understanding of the world's reliance upon oil and gas, examples we think about today may not be relevant tomorrow, as innovation and technological breakthroughs are happening at a very fast pace. But it *is* still relevant to note what today is 'unsubstitutable'. At the moment, there's no viable substitute for avgas, or aviation fuel, to power aircraft (although I note that various alternative power sources are undergoing active testing in small aircraft). We are a long way from completely getting rid of the petrol- or diesel-powered internal combustion engine, although a lot of work is going into doing so and timeframes are being established. As much as we are trying, it will be a long time before we have eradicated plastics from a whole universe of applications. Most chemicals come from hydrocarbons. The pharmaceutical industry is also reliant upon hydrocarbons for many applications. Perhaps ironically, given how some anti-fossil fuel activists also shun meat, just about all cheap and 'vegan' clothing comes from oil.
~~~~~~~~~~

In fact, approximately 20 per cent of hydrocarbon production goes into what's often referred to as petrochemicals, that is, oil and gas used for applications other than energy. So one-fifth of all oil and gas production is used for everyday applications which will take a very long time to substitute. To put this into context, in 2023, just over 96 million barrels of oil were produced per day.[12] That's about 35 billion barrels for the year. A barrel is 159 litres, so in 2023 the world produced 5.56 trillion litres of oil. The point is, 20 per cent of that, produced for non-energy, and thus non-burning purposes, is still an enormous amount of oil.

And yet this hasn't stopped a lot of activists demanding that funds, banks and others divest immediately from oil and gas companies. As I write this, a group of protesters are blocking roads in central London and spraying orange paint on things demanding that oil production cease immediately. It's not clear how they imagine the ambulances they keep disrupting are supposed to operate while we wait for alternative forms of motive power. Amusingly (although I appreciate it isn't amusing for the drivers stuck in the resulting gridlock), the glue with which these activists like to stick themselves to the roads of London is, of course, made from oil.

There's a demand that the oil and gas companies should reinvent themselves as renewable energy companies, and stop capital investment in oil and gas projects. Which leads to the same old problem: you can have oil and gas production in the hands of responsible, transparent public companies, or you can force them to divest their oil assets and drive oil and gas into the arms of opaque state-owned enterprises, unaccountable private equity funds and private companies domiciled in states with minimal or no reporting requirements and low or no ethical expectations.

The other problem is the reluctance to pay the economic cost of fossil fuel substitution, for example in relation to domestic heating, or to accept the inconvenience of EVs, particularly if you don't have a driveway.

What can the mining industry learn from this, and how can it avoid some of the problems that the oil and gas industry is dealing with?

The answer is: get better at explaining what mining does, demonstrate how it's essential and show that it's only getting more essential as we embrace a sustainable future. The industry also needs to deal with the 'outliers' who don't live up to the high standards society expects, and encourage/lobby for improvement. As discussed previously, the industry can more explicitly encourage investors and funds to support an orderly energy transition rather than rushing for the exit and leaving 'dirty' assets in the unaccountable hands of state-owned enterprises and opaque private companies. Miners can encourage investment in technology and innovation that will support the energy transition and more sustainable mining, while helping people understand that this technological innovation takes time and patience. Mining companies, universities and governments can support mining as a career where young miners can contribute to a more sustainable world. The industry can work with communities, NGOs, governments and charities to deal with illegal mining in a sensitive and practical way, work with and support those dealing with responsible sourcing and supply chain integrity and continue to work at stamping out corruption.

~~~~~~~~~~

If we are to have a responsible and sustainable mining industry, we need to have sufficient high-quality, well-trained people. Unfortunately, mining schools across the world are experiencing a sustained drop in undergraduate numbers. The metals demands of a sustainable future will require more miners, not fewer, so if this issue isn't resolved, it could bring the industry to a grinding halt.

This is a challenge for miners. I have mentioned the image problem of mining previously, and the hindrance that is to the recruitment of undergraduates to mining-related university courses. So how does the mining industry
~~~~~~~~~~

attract the miners of the future in the face of increasing need and an ageing workforce? This applies to both those with professional skills, such as geologists and mining engineers, and those with trade skills, to be the drillers, the roof bolters, the truck drivers, the digger and shovel operators and all the roles that actually get the ore out of the ground. Secondly, how do they attract the 'new' skills required by mining companies in IT and technology, to work on digitalization and all the facets of what's known variously as 'the digital mine', 'the mine of the future', 'smart mining' and so on? The underlying question is: how do miners convince enough people, and particularly young people with a career ahead of them, to consider a career in mining?

Firstly, the overarching need is to ensure people understand that a career in mining is valuable, acceptable, sustainable, diverse and laden with opportunity. Achieving recognition of the valuable and acceptable elements is completely connected with the premise of this book: that the mining industry needs to ensure the world understands mining's essential nature for everyday life and its contribution to sustainability and the energy transition. 'The world' includes prospective students and employees of the mining industry. The helpful part about this is that the industry can deal with two problems at once. By getting much better at explaining what miners actually do and why mining is essential, the industry should also help fix its recruitment challenge.

Frankly, if you believe in the sustainability agenda, if you care about the planet and the environment, if you want to do your bit, in a starkly practical way, to ensure a better future for everyone, then you should join the mining industry. It's reasonable to say, 'Save the planet, be a miner.'

The diversity and opportunity elements can also be addressed in a similar way. I believe that a lot of the problems that people perceive attach to mining as an industry apply in exactly the same way to mining as a career: it isn't a job for women; it's dirty; it may require spending time in unpleasant mine camps in almost inaccessible places; it often involves strenuous, repetitive manual labour; it's either boiling hot or freezing cold; it's boring; and there are few opportunities for advancement. These are the outdated and plainly wrong perceptions that a lot of people have in their minds when it comes to mining. If they were true, it would be pretty reasonable to accept that a lot of people would be turned off the idea of a career in the industry.

The reality is somewhat different. The breadth of skills required for modern mining and the diversity of roles on offer in the digital age have both increased significantly. The very successful campaigns to attract women to the industry, the fact that mechanized mining doesn't involve soot-covered men wielding pickaxes and the vast improvements in the quality of working conditions and accommodation are all emblematic of the change in the nature of working in mining. Finally, mining can be a genuinely rewarding career, with the opportunity to work across the world in a global industry and the opportunities for advancement in mining companies that place significant emphasis on career progression. All of these realities of the mining industry need to be highlighted to ensure prospective students of mining schools, and prospective employees of mining companies, have an up-to-date, accurate understanding of what working in the mining sector is really like.

It's not just the prospective students themselves who need to have these points highlighted. It's also important to explain the realities of mining to their parents, for whom the term 'mining' may bring to mind the unreconstructed and now almost completely disappeared legacy coal mining sector. Finally, and very importantly, these messages need to get to their school teachers. Many of them, to put it tactfully, have political views which can cause them to treat miners, and indeed much private enterprise, with unfounded but deeply ingrained suspicion. But they are also smart people. So when presented with the realities of modern mining, and particularly its essential contribution to implementing the sustainability agenda, I am sure they will come to appreciate the importance of encouraging their students to consider careers in the industry and to take courses at mining schools.

The breadth of skills and diversity of roles need some further comment. As we have noted previously, mining today requires IT skills and capabilities well beyond traditional mining-specific applications. It requires people who can work on the digitalization of mining and the development of things like remote operating centres; it requires process engineers who can think well outside the box; and it needs innovators who can apply disruptive thinking to traditional ways of operating.

The sector will be competing for this talent with many other industries, including the technology sector itself, but also manufacturing, transport and logistics and pharmaceuticals, among other areas. The miners can use this to their advantage, however. For example, working on the development of driverless haul trucks uses many of the same skills as working on the autonomous driving capabilities of domestic passenger cars, but is a whole lot more interesting, the vehicles are a lot bigger and it is of immediate, practical use. Another example is the development of augmented reality headsets so that remote technicians can assist in the diagnosis and real-time repair of machinery faults on mine sites. This is no different to virtual reality developments in other industries and even the gaming sector, just a whole lot bigger, a whole lot more real and a whole lot more exciting.

This competition for IT and technology students and graduates just adds to the importance of ensuring the world at large, and students and prospective employees in particular, have the right understanding of mining.

I noted above that dedicated mining schools across the world are having difficulty attracting students on to some of their courses. The mining industry really needs to get behind the efforts of these schools to attract sufficient students to enable them to continue. History tells us that once you lose these areas of academic specialization and expertise, it's very hard to recreate them.

So the industry as a whole needs to figure out its response to the employee attraction issue pretty quickly, and then needs to be quite robust in setting out its stall. The consequence of not doing so, at least in more highly skilled roles, is that the mining industry may run out of workers. It will be very hard to deliver the benefits of responsible, sustainable mining if the industry has no people.

~~~~~~~~~~

The purpose of this book is to consider what mining is, what it involves, why it's essential for sustainability and the energy transition and, finally, in the chapter we are just concluding, what sort of mining industry we want. The summary answer to that question is we want mining to be the very best it can be: responsible, sustainable and essential.
~~~~~~~~~~

8
Mining is essential for a sustainable future

We have, over the course of this book, established that mining is essential for a sustainable future. We have also explored what mining is, how it works and many of its challenges. These challenges, by their nature, can come across as a bit negative.

We mustn't make excuses for mining as an industry. It's certainly not always perfect; it's true that elements of it have a chequered history, and even today there are unfortunately mining companies who have little interest in being good corporate citizens and doing the right thing. But mining is fundamentally valuable, legitimate and needs to be respected as such. It deserves to be at the top table when the corporate world gathers for dinner.

So as we conclude, let's focus on the positive side by considering some of mining's contributions.

Mining enables our quality of life

The first chapter set out our total reliance upon the products of mining for our quality of life, and I won't repeat that here.

Some people suggest we only need mining to the extent we do so we can sustain our current energy-dependent, high-tech lifestyles. 'If only we could return to a simpler way of living,' they opine, 'our dependence upon metals would reduce.' Well I'm not sure that's true. We have been reliant upon metals for millennia. The nature of those metals may have changed, but the fundamental reliance upon metals has always been there. There's a reason why we use historical terms like 'Bronze Age' and 'Iron Age' for broad sweeps of history.

But there's a further concern with the 'return to the simple life' idea. And that is that for the vast majority of people, returning to a life without many of the conveniences, benefits and enjoyments that our metal dependence enables is just not something they have any appetite for. Yes, they want to live more sustainably, and, yes, they are very aware of the need to manage the environment better. So they are happy to make changes, and indeed that's already clear as people voluntarily seek to be more sustainable in the small things like reusing plastic and household recycling, up to the big things like their choice of motor vehicle and how they get around. However, they still want to be able to communicate easily, to travel conveniently and safely, to enjoy warm and comfortable homes and to enjoy a social life, as social beings. And another aspect of modern life that people rather want to hang on to is better health and the significant increase in life expectancy that goes along with it. And this quality of life, as we have said many times, is enabled by metals from mining, for which everyone should be thankful.

Mining enables sustainability

This is the point of this book and I don't need to repeat it here, but I include it in this list for completeness.

Mining underpins the economy and enables financial stability

The whole operation of the economy for both goods and services is held together by metals. Goods, as we have noted, are largely made from them, and services couldn't function without them. The economy's infrastructure which underpins everything is very metals intensive. But then I hear you ask: 'What about financial services?' Well, that's a fair question. The financial services industry can be summed up as the business of getting money from those who have it to those who need it. It is, firstly, very infrastructure intensive – all that IT, communications infrastructure and so on. But, perhaps more importantly, most financial transactions are underpinned by something tangible – be it trade in a commodity (often metal!), trade in real estate or loans secured against property. And, finally, the underpinning for the world's currencies for a very long time, and still the ultimate store of value, is gold. All that underpinning can only happen with a huge contribution from metal.

Mining facilitates all the other industries

Again, this statement encapsulates what we have talked about previously. The mining industry enables all the other industries by providing essential materials they need. Agriculture, medicine, pharmaceuticals, clothing, transport, communications, energy, housing, defence, construction and all the other industries owe their capacity to operate to the products of mining.

Mining enables international trade

We have noted mining's positive contribution to the economy and industry, and it follows that mining is thus also a contributor to international stability. And that's because mining enables international trade: the metal that powers global communication systems and global financial systems; and most importantly the metal used to manufacture all those ships, trains and aeroplanes moving commodities and finished goods across the globe. And it's very important to note that mining often happens in countries that don't have developed secondary industries, so the income from producing and trading metal is a significant contributor to those countries' wealth.

Mining contributes to national security

A primary goal for most countries, where it's feasible, is some measure of self-sufficiency. For a long time, many national governments didn't pay much attention to this aim, or it was lost under a wave of enthusiastic participation in globalization. The policy makers loved globalization, with all its perceived efficiencies, breaking down of barriers and buying into the principle that a steady inflow of affordable consumer goods equalled a happy, content and uncomplaining electorate. And it's important to say right now that I'm a firm believer in sensible globalization. A strong network of international trade can enhance national prosperity, lift millions of people out of poverty and contribute to global political stability.

But maintaining the ability to produce much of what you need domestically is also sensible, particularly items of a 'strategic' nature. And what do you make them from? Mostly metal. So when a global financial crisis, or a pandemic, or a war (either hot or cold) impacts materially upon international trade, having access to your own raw materials, and then having the capacity to process them and turn them into useful things, is extremely helpful. A domestic mining industry, preferably accompanied by a domestic manufacturing industry, is a contribution to national security. We flagged this up in more detail in chapter 6 in our discussion on critical minerals and resource security.

Finally, but very significantly on this topic, mining literally enables national security as the ships, aeroplanes, armoured vehicles, munitions and other equipment of national defence forces are made almost entirely from metals. Nations can't defend themselves without metal.

Mining provides opportunities

We have noted previously that mines are often in remote places. Consequently, it's quite common for mines to be the primary source of employment for adjacent communities. Mines provide stable and reliable employment where the only alternatives may be uncertain and seasonal agricultural jobs, government handouts or abject poverty. It's not just direct employment either. Mines provide opportunities for contractors, for local suppliers and for the providers of outsourced services such as security, cleaning, maintenance, deliveries and much more. Mining, across the world, is opening up opportunities for local communities.

Mining drives innovation

The mining sector is at the forefront of technological innovation. This innovation is in mining itself through the use of technology to improve efficiency and reduce material movements; in processing through innovative energy- and water-saving technology; and through digitalization of monitoring, analysis and control systems. Innovations in the rehabilitation of closed mines are also significant. In all these areas, mining is a leader in the development and application of innovative solutions. The results are also clear and should be celebrated. These include improved safety, reduced pollution and improved environmental stewardship, a significant contribution to sustainability and improved efficiency and productivity – getting the same result using less energy and fewer resources. It's important not to forget the flow-on effect, where innovations in the mining sector benefit adjacent industries, such as the development of remote operating centres, autonomous vehicles and industrial safety. As noted previously, it's important for miners to flag up their embrace of innovation: it's world class and should be acknowledged as such.

Mining brings joy

Almost all jewellery, and certainly all expensive jewellery, with the notable exception of pearls, is made from the products of mining. Gold, diamonds, silver, emeralds, platinum, rubies and every other gemstone you can think of are dug up. Think of all those wonderful gifts, for special occasions, anniversaries, birthdays and just to show you love someone: rings, necklaces, bracelets, earrings, pendants, cufflinks, tiaras, broaches and so on. And think of all the engagement rings! The crown jewels are made entirely of precious stones and precious metal (other than the ermine bit).

I acknowledge that artificial diamonds have become much better over recent years, and in fact those from reputable manufacturers are almost indistinguishable from the real thing to the naked eye. But, particularly at the high end, the real thing, a mined natural diamond, is still required. You may not be able to tell that it's fake, but you *know* it's fake, and that won't do! It's also worth noting that artificial diamonds use an astonishing amount of electricity to manufacture, as they require intense pressure and very high heat. So if you want to be sustainable, buy a real diamond, and enjoy it.

Given, therefore, that all this marvellous jewellery comes from metals and stones that start off under the ground, it makes me very happy to say: mining brings joy!

Mining has had, and continues to have, problems it needs to fix. Many of them are set out in this book. But next time you hear about some mining-related catastrophe, acknowledge the issue, but, please, remember mining's contributions.

~~~~~~~~~~

What are the key takeaways from this book?

Firstly, I trust by now you are satisfied with the book's basic premise: that mining is essential, and it's even more essential for a sustainable future.

Secondly, I hope that if you came to this book without much previous exposure to the ins and outs of mining, you will now have a much better understanding of the industry from the chapters on what mining is, what it does, what its challenges are and what responsible mining involves.

Thirdly, there are a few 'calls to action'. This is what needs to happen if mining is to provide the metals essential for a sustainable future, in a responsible and sustainable way.

Mining needs its key stakeholders to work sensibly and productively together. By 'key stakeholders', I am referring to the mining companies, their customers, their workers, communities, suppliers, shareholders and governments. In this respect, it's reasonable to expect that simple self-interest will make the miners, their customers and their suppliers work sensibly together. The miners, their workers and the communities impacted by (and benefiting from) their mines can also work together productively, as we have discussed in earlier chapters, and indeed responsible miners will ensure this happens.

The two stakeholders who perhaps need the most encouragement in this respect are governments and
~~~~~~~~~~

shareholders, more specifically institutional investors. The mining industry and the beneficiaries of mining (which is everyone) need governments to support this essential industry through sensible policies, consistent and even-handed regulation (including environmental regulation) and a stable tax and royalties regime. This will ensure an environment in which miners can invest with confidence for the long term that mining projects require.

Institutional investors similarly need to work sensibly with the mining industry, by thinking about both the long term and the big picture. I know these sound like clichés but when it comes to the mining industry, they are real. As we have noted many times in the preceding chapters, a mining project will take a decade or two to go from the feasibility study to starting production. Shareholders will then need a number of years of production before the 'pay-back period' is over and they start getting a return on their investment. So miners need investors to think for the long term. And then they need to think about the big picture. Producing metals for the energy transition is more than just digging up copper or lithium or nickel. It's helping ensure the energy transition actually results in sustainable outcomes. Maybe that means producing the cleanest (relatively speaking) thermal coal available. Maybe that means tempering their activism and recognizing that expanding copper mines in locations that they don't want is better than running out of electricity. Maybe it means supporting locally produced bulk metals, even if it looks suboptimal, because after factoring in all the shipping costs and the environmental impact of shipping commodities around the world, the local option is actually best. And maybe it means supporting listed companies hanging on to their operations producing unpopular commodities, to keep them accountable and transparent. We need miners to be able to depend upon the support of financiers and the capital markets if they are to produce the metals essential for sustainability and the energy transition.

My second 'call to action' is for miners and those who engage with them to really think about what sort of industry they want. Being essential doesn't give mining an excuse for poor strategy, poor performance and poor outcomes. What does 'mining for the future' look like? Clearly, it needs to be responsible, and it's easy to say 'it will be sustainable', but what does that look like in practice? I would suggest it means only developing mines that they know from the beginning can be both profitable and sustainable. It means using the best innovative thinking to develop a mine which uses the best mining and processing techniques to ensure the smallest possible environmental footprint while still producing a return for shareholders. It means a strategy, right from the beginning, that will ensure win-win outcomes for the mine and the communities around the mine, including a strategy for rehabilitation at the end of the mine's life. It means considering the whole value chain for the ore the mine is digging up: through processing to a concentrate or a metal, and then to mid- and end-users who will use the metal to produce sustainable and, hopefully, long-lasting goods that the world needs. And, finally, it means ensuring that the mine and the industry as a whole have the best possible people. This means actively supporting the mining schools who are educating the miners of the future for the mines of the future.

My last 'call to action' is to seek broader support for the mining industry. After all, it's essential. I neither want nor expect blind support. As we have discussed in this book, there are legacy issues the mining industry is dealing with, and even the most productive, sustainable and well-run mining operation will have room for improvement. But improvement where it's needed is much more likely with support and understanding from stakeholders and the broader public.

This broader support will include appreciation of the complexities and timeframes involved in getting new mining projects approved, financed and built. It will include an understanding of the risks miners face. It will include more considered decision making by investors in relation to their holdings in mining companies. It will include an appreciation for the work most miners are doing in the interests of sustainability both within and without the mine gate. It will include sensible and practical regulatory and taxation regimes. It will include miners, governments, NGOs and communities working together to improve the outcomes for those who live and work in and around mines. It will include support and assistance as miners deal with the insidious threats of corruption and illegal mining. And it will assist in attracting more people to work in the industry.

All these things are vital, because a strong, vibrant, growing mining sector is essential to meeting the ultimate challenge of achieving a sustainable future.

Appendix

Examples of our reliance upon metals and minerals

Below is a list of examples of our reliance upon mining. It is quite long but by no means exhaustive. I have arranged it by broad usage and then by metal. If a metal is a secondary metal, that is, the product of a process that combines other metals and minerals, I have shown that after the first time I mention the secondary metal.

Construction and infrastructure

To make glass for windows, solar panels or anything else, you need *silica sand* and *lime* from *limestone*, *magnesium* oxide and *aluminium* oxide.

To make *aluminium*, you need *bauxite*, which is then refined to make alumina, which is then smelted to make aluminium. Americans call aluminium 'aluminum' for some reason.

If the frames for your windows aren't made of timber, they will probably be made of *aluminium*, and then coated with a substance that is a combination of metals and petro-chemicals.

Many house roofs in the UK and elsewhere are made from *slate*.

The paint on those houses includes, among lots of other things, *silica*, *titanium* and *zinc*.

It would be very hard to have buildings more than a few storeys high without *steel* to hold them up and, particularly for buildings over about six storeys, to make the lift shafts, and for that matter the lifts themselves.

Everything attached to the ground uses *concrete* for foundations and structural elements. Just about all large buildings and infrastructure make extensive use of *steel* and *concrete*. Examples include bridges, road and rail cuttings and airport tarmacs. Look around you in any urban environment, everywhere there is concrete.

Concrete is made from *cement*, *aggregates* and *sand*. Cement is made with *lime*. Roads are made with *aggregates*, *sand* and other quarried construction materials (remember quarrying is just mining but simpler). They are then topped off with asphalt, which is basically oil.

Domestic applications

We could not enjoy Victoria sponge cakes or any other baking without *steel* or *aluminium* baking tins, and we would have no cooking of any sort without ovens made from *steel*, and *aluminium* and *copper* saucepans, frying pans and utensils.

My coffee machine is made from *steel*, *copper*, *brass* and *titanium*. My coffee grinder is made from *steel* and *copper*, plus ceramic grinding burrs, which are made from *clay*, hardened with *tungsten*. Clay is not a metal, but you do dig it up from the ground.

Electricity generation

There would be no electricity without *copper* for its generation and distribution.

There is still a dependence upon *thermal coal* for a significant portion of electricity generation, particularly in the developing world but also as back-up generation in most other places.

There would be even less electricity in many countries without *uranium* for nuclear power generation.

There would be no renewable electricity generation without *steel* for wind turbines, *glass* for photovoltaic cells, *copper* for transmission, *rare earth elements*, including *neodymium*, for the magnets and other REEs for the control systems.

To make steel, you need *iron ore* and *metallurgical coal*.

To harden steel, you add *molybdenum*, *vanadium*, *cobalt* and/or *tungsten*.

To make stainless steel, you need *iron ore*, *metallurgical coal*, *nickel* and *chromium*.

To galvanize steel, which is another way of protecting it from corrosion so it doesn't rust away, you need to dip it in *zinc*.

Gold

Gold gets its own subheading as it is perhaps the totemic metal. Gold continues as the ultimate store of value, over fifty years after the end of the gold standard.

Industry and agriculture

There would be no industrial production to make just about everything we use day to day without *steel*, *copper* and many other metals for the machinery, the controls, the factory buildings, the infrastructure, the tools and other ancillary equipment, and the vehicles to move everything around.

There would be very little agricultural production without *steel* for farm machinery and *potassium*, often called potash, for fertilizer.

IT, communication and appliances

There would be no electrical products, from small appliances to fridges to EVs, without *copper* for the wiring.

There would be no computers or other digital equipment without *copper* and many other metals, including *steel* or *aluminium* for the cases and frames.

Every computer, every digital piece of equipment, every vehicle, just about all modern machinery, and indeed anything with a chip in it has a circuit board. Everything is soldered to that circuit board with *tin*. Tin is the forgotten metal of the digital age. Nothing digital could work without it.

There would be no communications systems, both fixed and mobile, without *copper* for power and transmission, and *steel* and other metals for the infrastructure, including masts, towers and other structures.

The smartphone, tablet or anything with a touchscreen could not exist without *rare earth elements*, including *lanthanum* on the camera lenses.

Jewellery

There would be not much jewellery without *gold*, *silver*, *platinum*, *diamonds* and other *precious stones*.

Silver is used in jewellery and tableware, but most of it is actually used in industrial applications, including dentistry, semiconductors, soldering, water purification, LEDs (light-emitting diodes), nuclear reactors and many other things.

Medicine and pharmaceuticals

There could be no modern pharmaceuticals or medicines without sanitizable *stainless steel* equipment for research and manufacturing.

There are many metals used in medicines, including *iron*, *zinc*, *platinum*, *copper*, *lithium*, *potassium*, *magnesium* and *molybdenum*.

Surgical instruments use *stainless steel* together with *palladium* and other metals. Surgeons then use metals including *steel*, *cobalt*, *chromium* and *titanium* in the surgery itself. I write from experience: part of my left shoulder is titanium following an accident in the surf many years ago.

Transportation and mobility

There would be no vehicles of any sort without *steel*, *aluminium*, *copper*, *magnesium*, *zinc*, *nickel* and *platinum group metals*.

Modern internal combustion engines need *platinum group metals*, including *platinum*, *rhodium* and *palladium*, for their catalytic converters.

EVs need *cobalt*, *nickel*, *lanthanum*, *graphite*, *manganese* and *lithium* for their batteries, *aluminium* or steel for their bodies, *copper* for their wiring and electric motors and *rare earth elements* like *praseodymium*, *neodymium*, *samarium* and *dysprosium* for the magnets in their motors.

There would be no traditional car batteries without *lead*. In fact, lead has many other uses, including some piping and also in protection from radiation.

Hydrogen fuel cells used for hydrogen-powered vehicles need *platinum* as the catalyst to make electricity.

There would be no railways without *steel* for the rails and most of the infrastructure, *copper* for the power transmission and signalling and many other metals.

There would be no aeroplanes without *aluminium*, *copper*, *titanium*, some *steel* and lots of other metals. Some more recently introduced commercial aeroplanes use a lot of carbon fibre for their fuselages instead of much of the aluminium. Carbon fibre is essentially very high-tech plastic and comes from oil.

There would be very little international trade without *steel* because you wouldn't have any ships.

And a few other things just for fun . . .

Just about every statue in the world seems to be made from *bronze*. Bronze is made from *copper* and *tin* and is very corrosion resistant, so perfect for something you want to remain in a city square for hundreds of years. Bronze is also used for roofing.

As an avid user of an old-fashioned fountain pen for writing, I note that *rhodium*, in an alloy, is often used for pen nibs.

As a former trombone player, I appreciate that we need *brass* for musical instruments. Brass is made from *copper* and *zinc*, and is also used for locks, door furniture and in plumbing. As a current piano player, I also appreciate the cast *iron* frame of a piano, together with the strings made from high-tensile *steel* and *copper*. Whether my family and my neighbours always appreciate it is another matter!

~~~~~~~~~~

In summary, we use metals and minerals for everything. Mining, and I promise I am saying this for the last time, is essential.
~~~~~~~~~~

Notes

1 E.ON Next website, 'How Much Energy Does a Wind Turbine Generate?'
2 US Energy Information Administration, 'Electric Power Annual', October 2023.
3 International Atomic Energy Agency website, 'Nuclear Power Plant Safety'.
4 World Nuclear Association, Information Library, Nuclear Fuel Cycle, 'World Uranium Mining Production', 16 May 2024.
5 World Nuclear Association, Information Library, Facts and Figures, 'World Nuclear Power Reactors & Uranium Requirements', 23 October 2024.
6 EDF Energy, 'EDF Sets Out Progress at Hinkley Point C New Nuclear Power Station', 31 March 2017.
7 'Ireland Struggles to Consolidate Role as Data Centre Hub', *Financial Times*, 7 October 2024.
8 'Soaring Power Demand from AI and Crypto Pose Threat to Grid', *Financial Times*, 25 January 2024.
9 BHP, 'BHP Insights: How Copper Will Shape Our Future', 30 September 2024.
10 United Nations Trade and Development website data hub (unctadstat.unctad.org).
11 International Energy Agency website, 'Coal 2023, Supply'.
12 Statista, 'Global Oil Production 1998–2023', 25 September 2024.